Amateur Radio Novice Class License Study Guide-3rd Edition

by James Kyle, K5JKX,
Ken Sessions, K6MVH,
and Joseph J. Carr, K4IPV

THIRD EDITION

FIRST PRINTING

Printed in the United States of America

Library of Congress Cataloging in Publication Data

Kyle, James, writer on electronics.
Amateur radio novice class license study guide.—3rd Edition

Includes index.
1. Amateur radio stations. 2. Radio—Amateurs' manuals. 3. Radio—Examinations, questions, etc. I. Sessions, Ken W., joint author. II. Carr, Joseph J., joint author. III. Title.
TK9956.K92 1980 621.3841 80-21121
ISBN 0-8306-1273-4 (pbk.)

Contents

Preface

One of the major thrills in the world of radio is to engage in two-way communications. The amateur radio service is the only way one can legally talk across long distances, or to foreign countries, as a hobby. Citizen's Band operators are severely handicapped in this respect because of restrictions placed on the types of contacts which they are allowed to make.

The amateur radio operator is permitted to use a lot more power than the CB operator. The lowest class—Novice—operator in the ham bands is allowed to use up to 250 watts, which is 50 times the amount of power allowed the CBer. Higher grade amateur operators (i.e., General, Advanced, and Extra classes) are allowed to use 200 times the amount of power permitted the CBer (i.e., 1000 watts).

Amateur radio contacts vary from mundane chit-chat to serious public-service communication provided free. As this is being written, there is a story in the *Washington Post* newspaper of an amateur operator who relayed dozens of telephone calls to the relatives and friends of Americans trapped in a west African nation during a revolutionary upheaval. This amateur had been licensed only three months, yet found himself in the thick of things, bringing much welcomed mental relief to the families of those Americans. In times of natural disaster, such as hurricanes, earthquakes, and other calamities, amateur operators often provide the first communications between the stricken area and the outside world. The value of the individual amateur station is not overlooked by authorities, even in this era of supersophisticated public-communications systems. It has been amply demonstrated that single individuals, dedicated to the task, can get a simple amateur station back on the air much more rapidly than a major communications system. For the first few critical hours, or days, the burden of communications may well fall on the less vulnerable amateur operator. Many operators have stayed on duty during disasters for many long hours before commercial and emergency military communications facilities could take over. Even when some service is available through commercial channels, however, the main message traffic handled there will be official matters, leaving all of the "health and welfare" traffic to be handled via amateur radio (e.g., "Mom, Karen and I are OK, the flood didn't hit until. . ." Although officials must necessarily assign such traffic a low priority when commercial communications resources are limited, the amateur can handle a large volume of such traffic, to the relief of loved ones and friends of people stuck in the disaster zone.

In amateur communications you may talk to kings or kids, celebrities or working people, geniuses or dullards. The range of people found in amateur radio is immense, and most are not "electronics wizards." Looking in the *DX Callbook* (a directory of foreign amateur stations; another volume covers U.S. stations) you will find listed King Hussein of Jordan, several princes of the Saudi royal family, and, in numerous other countries, return addresses of "The Palace . . ." You will also find amateurs at 123 Main Street, Podunk Village, USA, and in the US Senate.

The newspaper article cited earlier claimed erroneously that amateurs had to pass tough examinations in electronic theory and "advanced mathematics." That is pure bunk. The exams are not all that difficult, even at the highest level. The Novice examination is simple enough that basic literacy is the most that is needed. There are numerous cases of youngsters in their early teens (one as young as 5!) passing the exams with high scores, and these are not child geniuses . . . just ordinary bright kids with the will to learn the material. It is well within the grasp of anyone who can read these words.

Amateur activities are many. Some just like to talk ("rag-chew"), some like to work in traffic nets, passing free radiograms as a public service. Some of the most skilled amateur operators fit this catagory. Others like to talk to only foreign countries (DX), and try to pile up large totals of countries contacted (there are, or have been, over 300 "countries" as defined for amateur-radio purposes). Still other amateur operators use the hobby as a vehicle for study and experimentation in electronics and certain related fields. Many of the practicing electronics engineers and technicians in industry got their start in amateur radio. Still others like to combine traffic handling with DX, and operate in networks such as the Halo Missionary Net (21,390 kHz daily at 1300 EST) or the Inter-American net (daily, following the Halo Net on the same frequency). The Halo net has regular check-ins from missionaries in Africa and South America. It is quite common to hear a Wycliffe Bible Translators Missionary working his amateur radio set from a village hundreds of miles up the Amazon River.

Amateur radio is an exciting hobby. In this book, we intend to provide all of the information that you will need in order to pass the Novice-class examination, and to get a head start on the General/Technician class license.

The Novice license is available to anyone, except the representative of a foreign government regardless of age, who does not already hold another amateur license. The examination for this license is deliberately kept as uncomplicated as possible. It consists of two parts; the first is a code test covering the sending and receiving of the international radiotelegraph code at 5 words per minute, and the second is a written test consisting of about 20 questions dealing with the most basic points of regulations and with the broad outlines of radio theory and operating technique.

All radio operation in this country (except for certain government operations) is regulated under authority granted by the Communications Act of 1934 and subsequent amendments. All examinations for radio operator licenses are conducted by FCC representatives, but in the case of the Novice and certain other amateur licenses, the "representative" is not an FCC employee. Instead, "volunteer examiners" administer the tests. A volunteer examiner must hold an amateur license of at least a General class.*

To prepare would-be amateurs, the FCC publishes a list of "study questions" and a number of firms publish "study guides" (even as we are, here). All of these guides are based on the FCC list of study questions; some of the published guides consist simply of the FCC list, together with direct answers to the specific questions appearing on it.

We don't work that way, since we feel rather strongly that memorizing the answers doesn't really teach anybody very much and the whole purpose of the test is to find out what you *know* rather than what you have memorized. Our "study course" is based

*Check with your nearest FCC Field Office for the latest information on the volunteer program.

on the FCC questions too, but we take a much broader view. In most cases, we paraphrase the original FCC question into a much broader one which covers all aspects of its particular subject. Naturally, this takes a lot more time and trouble—but the result is a deeper knowledge of the material, and much better chances of passing the examination on the first try.

If you already are familiar with some elementary electronics, then you might want to skip ahead to Chapters 8 and 9. If you can successfully answer more than 80 percent of the questions without "peeking," then it is likely that you can pass the FCC examination. You should study the FCC Rules and Regulations, Part 97, prior to taking the examination, and in fact, should maintain a current copy of these rules all the time that you are an amateur. Copies are available from the Superintendent of Documents, Washington, DC for a small fee. The FCC also publishes a booklet on interference that you would be wise to obtain.

No study guide can honestly promise that you will pass the examination, because the FCC will not release the actual questions (that would make it *too* easy!). The purpose of the examination is not to make amateur radio "exclusive," but to ensure that the operator knows enough to successfully control the radio transmitter so that interference to other stations is kept to a minimum. What we can promise, however, is that this study guide has been prepared according to the latest FCC examination syllabii. If you acquire enough knowledge to answer the questions in Chapters 8 and 9, therefore, you should have little trouble with the examination.

Joseph J. Carr, K4IPV

Chapter 1
Learning The Language

The starting point for any attempt to get into amateur radio is to learn the language. This is made a bit less than easy by the fact that you don't have just one new language to learn. We have counted at least six different languages—in addition to normal everyday English—which are necessary to the would-be operator.

Don't let this scare you off, though. All of us already speak several dozen languages, because each specialty has a language all its own. If you delight in doing your thing, then, whether you know it or not, you're involved with some language other than English—call it "mod" or "now" or whatever you like. And the six languages required for radio are all much closer to everyday English than is much of the dialog you hear on television.

To be more specific about it all, the six "additional languages" we intend to expose you to here are (1) electrical units of measurement such as volts, amperes, watts, hertz, picofarads, and the like; (2) formal radio technical terms such as choke, filter, amplifier, oscillator, key, antenna, etc.; (3) "amateurese," or amateur-radio jargon such as ham, landline, junior op, homebrew, final, gallon, and so on; (4) "FCCese," or the legal terms used in rules and regulations, such as notice, frequency bands, emissions, A1, and such words; (5) the Q-signals originally intended for radiotelegraph operation and now a part of ham jargon even in voice operation, such as QTII, QSL, QSB, QRM...; and finally (6) the "procedure signs" which have no official standing (with a single exception) but which are universally used in Morse-code operation, such as AR, K, R, DE, SK, and the like.

Obviously we cannot, in our limited space, teach you all there is to know about any one of these "languages," let alone all six of them. We can, however, show all that's necessary for the Novice examination for each of them, and, where practical, also point out the rules by which unfamiliar words of that particular language can be untangled.

ELECTRICAL UNITS OF MEASUREMENT

All electrical arts and sciences, including radio and electronics, require continual dealing with electrical quantities. Most of the quantities must be measured from time to time, and so we have a seemingly endless set of units of electrical quantities with which to measure them.

The basic quantities for electricity are usually considered to be volts, ohms, and amperes. Volts are the standard units for measurement of "electromotive force" or "potential," which is the electrical equivalent of the pressure in a gas or water pipe. That is, it's the force which "pushes" the electricity through. The word "volt" comes from the name of Allessandro Volta, an early-day electrical experimenter. Because it's one of the basic standards, the volt is defined in rather exotic terms, involving the attraction between two wires in a vacuum. For practical purposes, we know what a volt is by comparison with the national primary voltage standard, or a secondary standard which has itself been compared to the primary standard. Such a standard is the "Weston cell" which is used in physics laboratories for accurate voltage calibration.

An ampere is the standard unit for measurement of electrical current flow; that corresponds to the "gallons per minute" measurement made on a water or oil pipeline. The ampere is

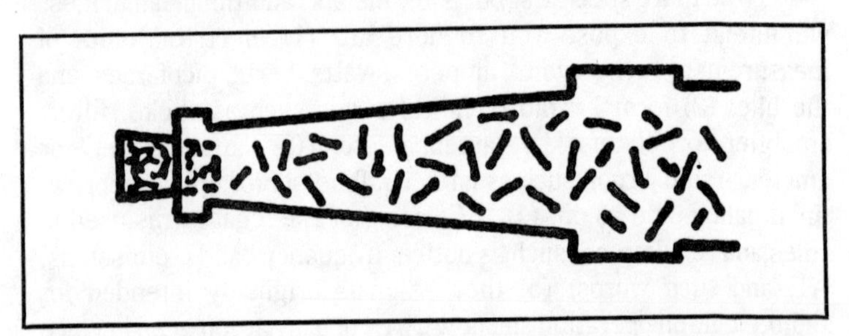

Fig. 1-1. Volts = potential, or "pressure" of electric force, as in an unconnected battery or like pressure in a stopped hose nozzle.

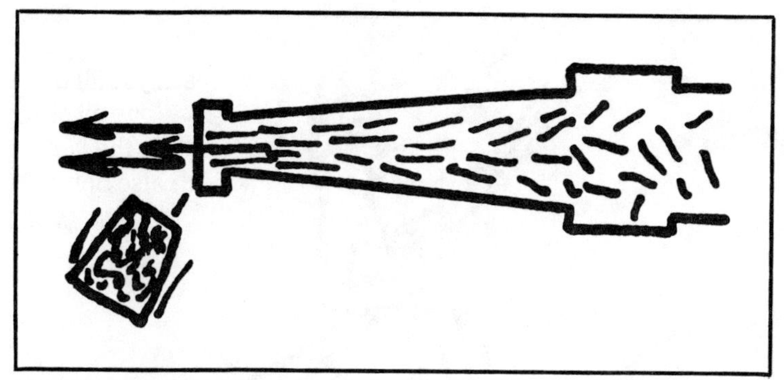

Fig. 1-2. Amperes = flow of current, with relation to time, as in gallons per minute from an unstoppered hose.

defined internationally, like the volt, in terms of physical measurements. It is also defined as the movement of one "coulomb" of electric charge (this quantity is similar to a "gallon" of liquid) past a given point in one second of time. The important point here is that amperes measure current flow, and they measure it with reference to *time*. There's no such thing as an "instantaneous current," although often it's convenient for us to act as if one could exist. If the electricity is not moving current is not flowing.

An ohm is the standard unit for measurement of electrical resistance. Resistance is that property of all materials which "resists" the flow of electric current. This unit is named for a

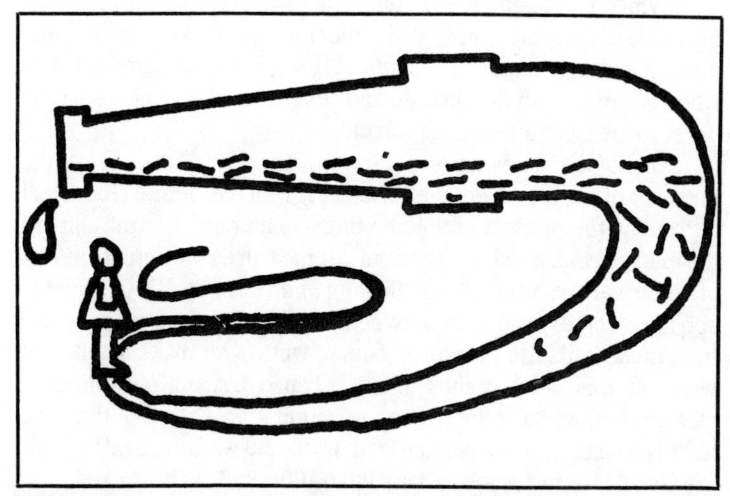

Fig. 1-3. Ohms = resistance, restricting the flow of current; standing on the hose decreases the flow.

Fig. 1-4. Watts = amount of power, or work done, as in horsepower. Pressure times rate of flow indicates the energy used or the work done.

German schoolteacher, Georg Simon Ohm, who discovered the constant relationship between volts, ohms, and amperes which we know today as Ohm's Law—and which we'll go into in a later chapter.

While volts, amperes, and ohms are basic to all electrical measurements, they are also sufficient to make any desired direct-current (dc) measurements. A few other units come in handy when working with dc, though, and several more are necessary in order to measure alternating current (ac).

One of the handy units is the watt, named for James Watt, the inventor of the steam engine. This unit, which is defined for direct current as the product of voltage times amperage, is a measure of the energy consumed in any circuit. It measures the actual amount of electrical force *and* charge flowing in the circuit. It's possible to set up a correspondence between watts and horsepower, since both these units are measurements of work. Our major interest in watts at this point comes about because transmitter power is measured in watts delivered to the output stages. In CB, the legal limit is 5 watts; for the Novice ham, up to 250 watts is legal, and for the General or higher class ham, up to 1000 watts may be used.

When we go to measure alternating current, it's not enough to measure its pressure and the current flow. We also have to know how frequently the current changes polarity, or "alternates," because much of an ac circuit's behavior depends upon its operating frequency. The quantity we call *frequency* is measured in units called *hertz* (abbreviated Hz), which are named for Heinrich Hertz, the first man to be recognized as having demonstrated wireless transmission and reception of energy. Until hertz was adopted as the official unit of frequency, all U.S. frequency measurements were made in "cycles per second" which was abbreviated "cps" or "c/s", and often shortened to simply "cycles." The relationship is 1 Hz = 1 cps.

Many electrical components have their own individual properties, such as the resistance of a resistor (which we already saw has its own measurement unit, the ohm).

Among the more important such properties are the capacitance of a capacitor and the inductance of a coil. Capacitance is the

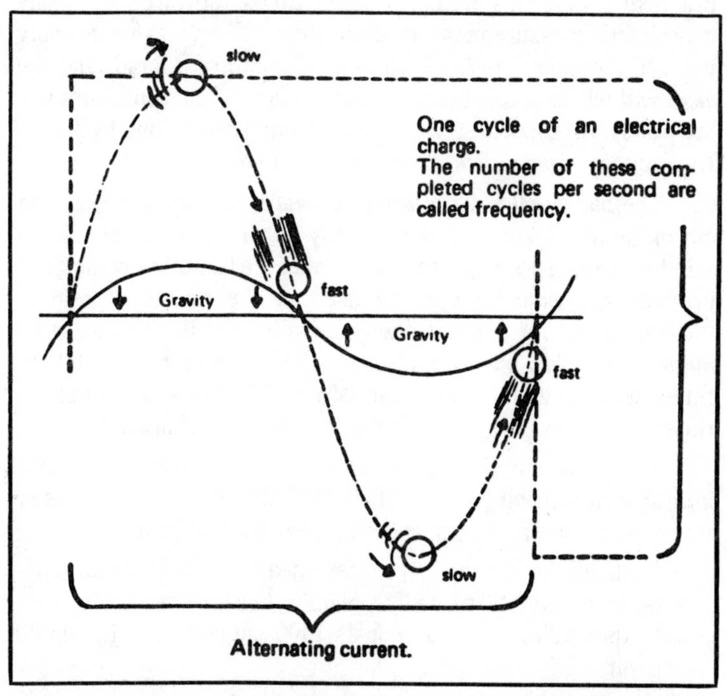

Fig. 1-5. An electrical charge in alternating current acts much like an item thrown into the air which starts quickly slows down at the top of the arc (where the charge is dead) then gains speed as it falls back to earth and is immediately shot from earth again (this time with poles reversed) slowing down as the charge dies and falls back to the ground plane.

ability to store an electric charge, which amounts to voltage without current. Static electricity, for example, which shocks us in the wintertime, is accumulated by the capacitance of our bodies. We don't feel it as long as it's stored because no current flows, but when we touch something to which the charge can flow, we get shocked.

Capacitance is measured, officially, in units called *farads*. The name honors Michael Faraday, one of the most illustrious of early electrical experimenters and the inventor of the dynamo. One farad is the capacitance which would store one coulomb of charge at a potential of one volt. (We met the coulomb briefly a little earlier—it is defined as the charge of 6,280,000,000,000,000,000 electrons if you're interested, but this unit is used primarily by physicists and has little or no place in amateur radio measurements.)

As it happens, the *farad* is about a million times too big for any practical use, and a million million times too large for many capacitance measurements made in radio. The units we actually use are the *microfarad*, which is a millionth of a farad, and the *picofarad* which is a millionth of a microfarad. The capacitors you are likely to encounter may range from several hundred microfarads capacitance down to a picofarad or so.

The names *microfarad* and *picofarad* were not simply pulled out of the air. All electrical units may be multiplied or divided in size by putting special prefixes in front of the basic unit name. The prefixes are essentially those used in the metric system of measurement, but the only ones generally used in radio are *giga, mega, kilo, milli-, micro-,* and *pico-*. *Giga* means a unit one billion times as large as the basic unit. *Mega* indicates a unit a million times as large as the basic. *Kilo* means a thousand times as large.

The remaining three prefixes mean smaller than basic units: *milli* indicates a unit one one-thousandth the size of the basic one, *micro* means one-millionth, and *pico* means one-billionth.

Voltage, for example, is often measured in *kilo*volts, *milli*volts, or *micro*volts, as well as plain ordinary *volts*. One kilovolt equals 1000 volts. One volt equals 1000 millivolts, or 1,000,000 microvolts.

The system of prefixes to indicate size of units may seem a bit cumbersome at first, but it actually makes things much easier because the actual numbers involved in any measurement can be kept small simply by choosing the appropriate unit name. The high voltage in a TV picturetube circuit averages about 25,000 volts—but it's much easier to remember that it's 25 kilovolts.

The same system of prefixes, incidentally, is coming into more general use. For instance, the force of nuclear explosions is measured in "*kilo*tons" or "*mega*tons" of TNT, meaning thousands or millions of tons respectively. And the cost of some military weapons systems is sometimes quoted in "*mega*bucks."

Now that we've seen how the prefix system works, let's go back to our standard basic units:

Inductance is the ability to store energy in a magnetic field. It's similar in many ways to capacitance, which stores electric charge—but exactly opposite in many others. Physicists say that inductance and capacitance are "duals" of each other, meaning that they have many similarities and certain critical opposites. Another familiar pair of duals are voltage and current.

Notice the significance of the capital letters when used as an abbreviation: MV means Megavolt; mV stands for millivolt. Table 1-1 shows the most common abbreviations used in electronics.

The unit of inductance is the *henry*, named for Joseph Henry, inventor of the relay and the telegraph receiver (Morse's critical contribution to the telegraph was to invent a code and establish a

Table 1-1. Prefixes and Abbreviations for Units Commonly Used in Electronics. Blank Spaces in the Example Table Mean that these Values are not Common. For Instance, Millifarad (mF) is not Used; the Value Could be Expressed as 0.001 Farad, but More Often is Written as 1,000μF.

M = Millions; X 1,000,000; (X 10^{6}); Mega

k = thousands; X 1,000; (X 10^{3}); kilo

d = tenths; X 0.1; X (1U^{-1}); deci

m = thousandths; X 0,001; (X 10^{-3}); milli

μ = millionths; X 0.000001; (X 10^{-6}); micro

n = thousandths of a millionth; X 0.000000001; (X 10^{-9}); nano

p = millionth of a millionth; X 0.000000000001; (X 10^{-12}); pico

Examples:

V	MV	kV	—	mV	μV	—	—	volts (electromotive force)
W	MW	kW	—	mW	μW	nW	pW	watts (power)
F	—	—	—	—	μF	nF	pF	farads (capacitance)
Ω	MΩ	kΩ	—	—	—	—	—	ohms (resistance)
H	—	—	—	mH	μH	nH	—	henrys (inductance)
B	—	—	dB	—	—	—	—	bels (relative strength or loudness)
Hz	MHz	kHz	—	—	—	—	—	hertz (frequency; 1 Hz = 1 cycle per second)

working system). Henry was the first serious American student of electricity—with the exception of Ben Franklin.

The henry is defined in terms of volts and amperes, as being that inductance in which a voltage of 1 volt will be induced by a current change of 1 ampere per second. This is a practical size of unit for audio coils and filter chokes; for radio-frequency work, the units millihenry and microhenry are used.

FORMAL TECHNICAL TERMS

By far the most complex of our six languages is that of "formal technical terms." This language contains several thousand words (by contrast, the total vocabulary of the average adult is only about 5000 words, and even Shakespeare employed only about 25,000 words), and obviously we're not about to introduce you to all of them here.

Throughout this volume we'll be continually introducing you to various parts of this language, and since you're probably already interested in radio and electronics, you've already met some of the more common parts of the formal technical language.

In fact, all we are going to do at this point concerning this one of our six languages is to tell you that it exists, that you will be exposed to it and will keep learning it as long as you are interested in electronics, and mention several dictionaries and encyclopedias which you can obtain if you want a reference book to it.

Actually, most of us who are today deeply involved in amateur radio and allied fields learned the language of formal technical terms without the aid of dictionaries, simply by reading all we could about electronics. This is approximately the same way a baby learns its native language—by using it. And it works rather well. If you continue with this study course, you needn't worry too much about the formal technical terms, because we're going to make every effort to introduce them to you gradually and clearly.

Amateurese

Every specialty has jargon. The computer engineer speaks of *bits, words, bytes, memory, clocks*, and the like; the race-car driver's conversation may include *drift* or *wheelstand*. Amateur radio is no exception. And, like all such jargon, that of amateur radio is usually colorful.

One of the most widely known words of the amateur's jargon is *ham*, meaning amateur radio operator or merely amateur (as in *ham radio*). Another is *ticket*, meaning an amateur radio operator license. This course is intended to help you get your Novice-class amateur license; a ham would say it's intended to help you get your Novice ticket.

No knowledge of amateurese is necessary to get the Novice ticket, so far as the FCC is concerned. The examinations are written in straightforward (at least so far as any government agency can be) language and no ham jargon is used. It will help, though, because it makes it easier for you to read the articles in publications intended for ham readers. What's more, it will help us present you the necessary information because frequently a short expression in ham jargon can take the place of a lengthy phrase in engineeringese words—and between engineeringese and amateurese, the amateurese is far easier to comprehend for most folk.

Like the vocabulary of formal technical terms, you'll pick up most of the ham expressions automatically along the way. We'll define here only a few of the expressions, chosen more or less at random, to give you an idea of how the jargon has been built up over the years:

Landline—a telephone or telephone circuit, so named because it depends on wire lines run across the land, rather than on radio waves traveling "in the air" (or at least, it used to. Now, many long-distance "landline" circuits use microwave relay instead of actual land lines, but the term has stuck.)

Junior Op or Harmonic—either of these may be used to refer to the children of a ham. *Junior op* expresses the hope that the child will also become a ham, while harmonic is a worn-out pun on the technical meaning of "harmonic" as a multiple of a fundamental frequency, and children are the result of multiplication.

Homebrew—equipment designed and built by the ham himself, rather than being "store-bought" or built from a kit. Some hams would have you believe that homebrew equipment is the only truly amateur way to travel; others consider all homebrew efforts a waste of time and material and insist that commercial equipment is the only kind worth having. The truth is somewhere between these two extremes; a good homebrew job is a work of beauty and something of which anyone can be justly proud, but many items of homebrewed equipment turn out to be, to put it mildly, junk. It all depends upon the ability and craftsmanship of the homebrewer.

Final—technically, the final stage of any electronic circuit, but in ham jargon this word means the final radio frequency power amplifier of a transmitter. That is, the stage which delivers the power to the antenna.

Gallon—a 1000-watt transmitter or amplifier, which is the maximum power permitted a ham operator.

Novice Gallon—a 250-watt transmitter. Derived from the "gallon" or "full gallon," since 250 watts is the maximum legal limit on a Novice station's power.

Bug—semiautomatic telegraph key, which sends dits automatically but does not form the dahs. This term evolved from the trade name of one of the first such instruments, the "Lightning Bug" key.

TO Keyer—a popular electronic bug, which forms both dits and dahs automatically. The "TO" comes from the call of W9TO, who invented the circuit.

Many words in amateurese have been picked up from the Morse-code abbreviations and "procedure signs" originally developed, like the Q-signals, for radiotelegraph communication. Some of these include *lid* for poor operator, *DX* for distance, *tnx* for thanks, etc. We'll look at more of these a little later.

FCCese

Like most government agencies, the FCC has developed its own favorite phrases through the years of its history. In addition, the Rules and Regulations which govern operation of all FCC-licensed radio stations *are* legal documents, written by and for attorneys as a general rule. The net result is that "the book" is exceptionally precise about what is and is not permitted—but its precision is expressed in words which aren't particularly likely to carry much meaning to the average radio operator.

For instance, let's look at section 97.112, from the regulations governing the amateur service: "An amateur station shall not be used to transmit or receive messages for hire, nor for communication for material compensation, direct or indirect, paid or promised."

It wouldn't serve most of the purposes for which the rule book was written to say instead, "You can't be paid in any way for sending or receiving a message over a ham station." ...but it would mean the same thing.

The purpose for all the extra words in the official version is to attempt to close any possible loopholes through which a violator might try to avoid the penalties of rule-breaking. To cite some parts from our example, "direct or indirect" means that it's just as illegal to accept a cup of coffee for delivering a message as it is to charge $500—should the commission want to enforce the rule that strictly. And "paid or promised" means that they need show only that you were promised payment; your claim that you never received it is no defense.

Not all the loophole-closing is for the government's convenience. It's often said, for instance, that it's illegal to make money from ham radio if you are yourself a ham—but the regulation says only that you cannot "transmit or receive messages" or use your station "for communication" and be paid. There's no prohibition at all against charging for the repair of a ham station, or its installation, or even being paid for holding office in a ham radio organization!

While we've picked apart section 97.112 here as an example, the same is true of most other parts of the regulations. We chose 97.112 only because it is short and offers examples of most of the main differences between "FCCese" and ordinary English.

Some of the "favorite phrases" you'll come across frequently in the regulations include:

■ *notice*—an official notice from the FCC to a licensee, normally informing him of some alleged violation of rules.

■ *emissions*—radio signals transmitted from a station; the phrase includes all signals transmitted rather than just the signals which you intend to transmit, and is often used in the term *spurious emissions,* meaning undesired signals.

■ *frequency bands*—the bands of frequencies assigned to one or another of the radio service.

One type of term which you'll come across time and again in all official documents, and often in plain technical reading as well, is due to the CCIR (international radio consultative committee; the official name of the committee is in the French language, which explains the transposed order of initials in the abbreviation). That is the coding for "type of transmission" or "emission type" such as A1, A3, F1, F5, etc.

All types of radio signals have been divided into three classes of modulation-AM, FM, and pulse (we'll find out more about these types of modulation in a future installment—only their names are important at this stage). The letter in the emission-type code tells what type of modulation is involved, with "A" indicating AM, "F" for FM, and "P" for pulse.

In addition, six classes of signal have been set up numbered 0 through 5, to indicate the purpose of the modulation. "0" indicates no modulation, "1" indicates keyed telegraphy, "2" is tone-modulated telegraphy, "3" is radiotelephone or voice transmission, "4" is facsimile, and "5" is television. Besides these six classes, a special "composite" class with the number "9" to cover all cases which fail to fit into one of the regular six classes.

A signal coded as *A1*, then, is radiotelegraph using on-off keying of an AM signal. A code of *F1* indicates telegraphy with an FM signal, or frequency-shift keying. *A3* is normal AM voice, and *F3* is FM voice. *A5* would be AM television. Normal AM broadcast radio is type A3, FM broadcast is F3, and commercial television is A5F3 because the picture is A5 and the sound is F3.

Other numbers and letters are also added to this code at times to describe additional signal characteristics, but for the Novice examination they are not necessary.

Q-Signals

One of the features of a ham's conversation which immediately identifies his hobby is the liberal use of Q-signal abbreviations sprinkled throughout his speech.

"I was in QSO with a QRP op last night," he may say. "At my QTH the QRN was terrible, and QRM was taking me out at his end of the line. We tried to QSY, but that was no help, so we finally had to go QRT: but he promised to QSL."

This may sound exaggerated; if it does, you've never been to a ham-radio club meeting. Fortunately not all hams talk this way—but the FCC expects a would-be Novice operator to know the more common Q-signals and their meanings, because they do have a couple of very important purposes.

While one of the reasons often cited for development of the Q-signals was that of convenience—it's much simpler to send "QRN" by a telegraph key than it is to spell out "I am being troubled by static"—that's not the most important purpose of the Q-signals. After all, the "procedure signs" we're going to examine next share this quality of speed and convenience.

The main reason why the Q-signals are important is that they are an internationally recognized set of phrases which mean the same to all radio services and to all operators regardless of language. An American ham and a German may not know a word of each other's languages—but when one taps out "QRZ?" the other knows that he is being asked "Who is calling me?" and he can reply, "QRZ K4IPV" to mean "You are being called by K4IPV." In other words, the Q-signals form a world-wide common language for radio operators which transcends the national languages of each.

The Q-signals are all combinations of three letters, and the first letter is always "Q" to indicate that this is a Q-signal. If only the three letters are sent, the signal is a statement. If the three letters are followed by a "?", it is a question. With all possible combinations of the second and third letters, we have a possibility of 676 different Q-signals. Adding the question form doubles this number, so it would be possible to have 1,352 Q-signals—676 of them being questions and 676 being statements.

Actually, only a fraction of this number have ever been officially recognized, and only a handful of the recognized ones are in general use by hams.

Two Q-signals which begin with "Q" but are not officially recognized are also used by hams. One is *QST,* which has no question form, and means the message which follows it is a broadcast to all amateurs. The other is *QRRR*, which is sometimes called the "amateur's SOS." Both are sponsored by the ARRL, and, as we said, are not officially recognized. In an emergency situation, the *official* emergency signal is probably preferable—SOS for telegraphy, and "mayday" for voice communcation.

The following list of Q-signals includes almost all of those in use by hams, with the more frequently used ones at the head of the list:

QRM—Are you being interfered with? I am being interfered with.

QRN—Are you troubled by static? I am troubled by static.

QSO—Can you communicate with...? I can communicate with...

(The ellipsis—three periods in a row—indicates that the sign of a third station can follow the Q-signal and that callsign takes the place of the ellipsis in the question or statement. Many Q-signals have this option.)

QRT—Shall I stop sending? Stop sending.

QTH—What is your location? My location is...

QSY—Shall I change frequency? Change frequency to...

QSL—Do you acknowledge receipt? Receipt acknowledged.

QRZ—Who is calling me? You are being called by...

QRX—When will you call me again? I will call you again at (time) on (frequency). (Often used without options to mean simply "Wait a moment and I'll be right back.")

QSA—What is the strength of my signals? Your signals are (1 for scarcely perceptible, 2 for weak, 3 for fair, 4 for good, 5 for very good).

QSB—Are my signals fading? Your signals are fading.

QRS—Shall I send more slowly? Send slower (...words per minute).

QRQ—Shall I send faster? Send faster (... words per minute).

QRO—Need I increase power? Increase your power.

QRP—Need I decrease power? Decrease power.

And one more very unofficial Q-signal, used particularly by hams with humor to their closer friends, and referring to their telegraph keying habits:

QLF—Are you sending with your left foot? Try the left foot next transmission.

Procedure Signs and Abbreviations

In addition to the Q-signals, which have official international recognition, and the amateurese jargon of abbreviations, there are quite a few "procedure signs" used in radiotelegraph communication—and several of them show up often enough in print to require some knowledge of them.

Since the procedure signs have little or no official standing, it's advisable to be cautious in their use. They may leave the other fellow more confused than communicated with. If used with care, though, they can shorten either a telegraph conversation or a voice communication with no loss of information.

Some of the procedure signs apply only to telegraphy; in general, these are signs chosen only to minimize keying effort. One example of such a sign is *es* to mean *and*; sending the *es* characters requires only a single dit for the E and a series of three dits for the S, where the A, N, and D of *and* would require a total of four dits and three dahs distributed among the letters.

Another such example is *hi*, used to indicate laughter because the telegraph key cannot capture your chuckle at the other fellow's joke. Again, only six keystrokes are required.

Not all telegraph-only signs are chosen especially to reduce keying effort. Some substitute for whole phrases.

One such is the *de* required by regulations in the calling sequence between two stations, to separate the sign of the station being called from that of the station doing the calling. The *de* translates roughly as "this is." Using voice, one might say "W6FSM, this is K4IPV" and the same sequence in telegraphy would be "W6FSM de K4IPV." These forms, incidentally, are specified by FCC rules; you cannot legally reverse the order and say "K4IPV calling W6FSM."

Similar to *de* but not specified in the rule book, are the signals *AR, K, KN, SK,* and *CL.* The *AR, KN,* and *SK* characters are sent as one character rather than as two separate letters—*AR* is didahdidahdit, rather than didah (space) didahdit.

Meanings of these procedure signs are as follows:

AR—identifies "end of transmission." Corresponds to "over" as used in voice communication, to signal other stations that you are through and he can reply now.

K—go ahead. Similar to *AR*, but *K* is used after communication has been established where *AR* usually indicates that you are going to tune around to look for a reply.

KN—"go ahead" for the specific station you are communicating with, and "keep out" for all others.

SK—end of session; indicates that you are ready to end communication with this station.

CL—closing station. This is sent as two separate characters rather than being run together into one. Used only when no further listening is to be done, to indicate to everyone else that it's a waste of time to call you now.

Other procedure signs include:

CQ—general call to any station.

V—test in progress. During on-the-air tests of a station's equipment, a continuous series of V's is usually sent.

R—same as "roger" in voice communication, to indicate that all of the previous transmission was received correctly.

Closely allied to the procedure signs are the CW or telegraph abbreviations, which are often used in voice work as well. Some of these include:

OM—old man, referring to any male amateur.

OT—oldtimer, meaning anyone you've talked with before.

YL—young lady or girl friend.

YF—wife (say the abbreviation and you'll see where it came from). Many operators use *XYL* for wife too—from ex-*YL*.

GM—good morning.

GA—go ahead.

CUL—see you later.

CW—continuous wave, or telegraphy.

BK—break. Used to interrupt QSO in progress.

DX—distant station. May mean in next state, or on next continent, depending on operator's experience and interests. "Working *DX*" is one of the major thrills of ham radio, in the opinion of many operators. Others shun it and go QRT when *DX* is heavy.

Rig—station equipment. Often only the transmitter.

HR—here.

TNX—thanks.

TVI—television interference (dread subject).

UR—your.

VY—very.

WUD—would.

WX—weather. Many words are abbreviated to an initial followed by X. Others include transmitter (*TX*), receiver (*RX*), and of course, *DX*.

73—best regards.

88—love and kisses.

Now that we have met the six languages of ham radio, we can start learning them by using them, and learning the theory and rules necessary for the Novice ticket at the same time. Thus, the next chapter delves into the basics of electrical theory to show why batteries, resistors, capacitors, and other components work.

RST SYSTEM

The R-S-T system is used by amateurs to offer semi-objective information about each others signals. The letters R-S-T mean "readability-strength-tone." The readability refers mostly to the ease of copying of the signal through the interference, although for radiotelephone it might also be an indication of the quality of the modulation. Some amateurs include such factors as the hardness or softness of the radiotelegraph keying, or the ability of the operator to form morse code characters (i.e. the operator's "fist"). The strength ("S") is a scale of 1 to 9, and refers subjectively to how strong the signal is at the receiving location. The tone refers to the character of the emitted radiotelegraph signal, and is deleted for radiotelephone (in which case the RST system becomes an RS system). A perfect signal will be 559 for CW and 59 for phone.

Readability

1. Unreadable
2. Barely readable - some words distinguishable
3. Readable with some difficulty
4. Readable with practically no difficulty
5. Perfectly readable

Signal Strength

1. Faint signals, barely perceptible
2. Very weak signals
3. Weak signals
4. Fair signals
5. Fairly good signals

6. Good signals
7. Moderately strong signals
8. Strong signals
9. Extremely strong signals

Tone

1. Extremely rough, hissing, note
2. Very rough AC note, no trace of musicality
3. Rough low-pitched AC note, slightly musical
4. Rather rough low pitched note, slightly musical
5. Musically modulated note
6. Modulated note, slight trace of whistle
7. Near DC note, smooth ripple
8. Good DC note, just a trace of ripple
9. Purest DC note

Chapter 2
First Comes Theory

Because the primary reason for issuing amateur-radio licenses in the first place is to provide a pool of "well-trained and experienced operators" in case of national emergencies, the FCC requires that every applicant for an amateur license prove his knowledge of both the theoretical and the practical aspects of radio.

The complexity of the examination depends, of course, upon the class of license sought. For the Novice license, only a minimum amount of basic theory is required. In fact, it's just enough to assure that the licensee will be able to operate his equipment within the limits set by the regulations! However, when it's all new to you, even this minimum theory may appear highly complicated.

We'll try to ease you over the hurdle while we trace through the elementary building blocks of all electrical theory. We'll keep the math out of it as far as possible, although a touch of algebra (the kind taught in junior-high-school "new math" these days) is just about inescapable.

THE BUILDING BLOCKS

It's a good thing that none of the FCC examinations contain the question "What is electricity?" because that's one nobody has yet been able to answer accurately. The best that our sharpest brains can do is to provide a number of *theories* which appear to fit all observed facts.

Of these theories, the one which currently is used by almost all physicists is the "electron theory." It explains not only electricity, but all physical matter, in terms of "atoms" which are in turn composed of "atomic particles." In this theory, three types of atomic particles are recognized; they are the "electron," the "proton," and the "neutron" (Fig. 2-1).

The theory is so completely accepted that you may be surprised by our calling it a "theory"—quite possibly you have been taught that it is "fact." It is, however, not yet proved beyond all doubt, and so must be considered as merely "the best explanation to date."

The electron (from here on, we won't hedge by inserting the phrase "according to the theory") is a tiny particle of energy or matter, with a "negative" electric charge. It is, in fact, usually called "the unit of electric charge." The neutron is much heavier than the electron, and is electrically neutral. The proton is of the same mass as the neutron, but has a positive charge. (Some variations of the theory consider the neutron to be a bonded pair consisting of one electron and one proton; others call it a separate particle. Take your choice.)

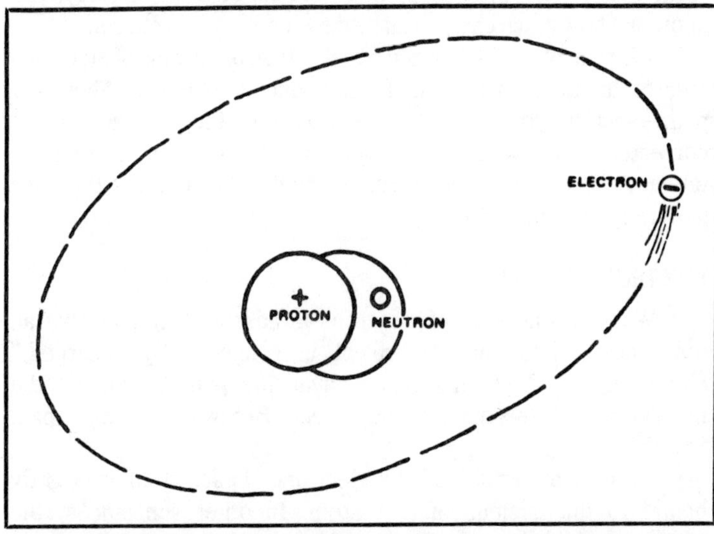

Fig. 2-1. According to present atomic theories, all matter is composed of molecules, which are in turn made up of atoms. An atom is the smallest particle of a chemical element capable of retaining the properties of that element; but atoms in their turn are composed of atomic particles called neutrons, protons, and electrons. The neutrons and protons make up the "nucleus" of the atom, while the electrons are in "orbital shells" at a relatively great distance from the nucleus. Electrical effects are due entirely to action of electrons. Shown here is a not-to-scale sketch of a helium atom, the simplest one which contains all three types of particles; the hydrogen atom is similar but has no neutron.

Since all matter is made up of atoms in some kind of structure, all matter contains electrons and protons. What we call "electricity" is an imbalance between the number of electrons and the number of protons in any given item.

One of the observed facts which helped lead to the electron theory was that objects with like electrical charges repelled each other, while those with opposite charges attracted each other strongly (Fig. 2-2).

Because of this attraction, in most objects the electrons and protons balance each other and the object itself has no charge, either positive or negative. The electrons and protons inside the atoms of the object remain separated from each other, despite the attraction forces, because of other forces in the atomic structure.

If, for any reason, an object either loses some of its own electrons, or gains excess electrons from somewhere else, it is said to be charged. One way to gain or lose electrons it to rub them off the surface by friction; this happens when wool slides across certain types of plastic—and if you have ever suffered an auto "seatcover shock" upon stepping from a car on a cold dry day, you know just how much charge can be developed by this means.

When an object is charged, its shortage or surplus of electrons creates an imbalance with all surrounding objects. When any path—such as your body—offers a path for the imbalance to be corrected by a flow of electrons one way or the other, that path is taken. A flow of electrons through any substance is the thing we generally call "electricity."

ELECTRONS IN MOTION

When the number of electrons in an object is such that an imbalance exists and the object is consequently "charged," electrons will flow through any available path to correct the imbalance and "discharge" the object. But what makes a path "available"?

In some materials, all the electrons of each atom are tightly bound to the nucleus of that atom. In other substances, the electrons which form the "outer orbit" (farthest from the nucleus) are not so tightly bound and may move from one atom to another in a haphazard manner. Electrons which are able to move in this fashion are called "free electrons"; the atomic structure of any material is the chief factor determining whether it has free electrons at low temperatures and many free electrons at higher temperatures.

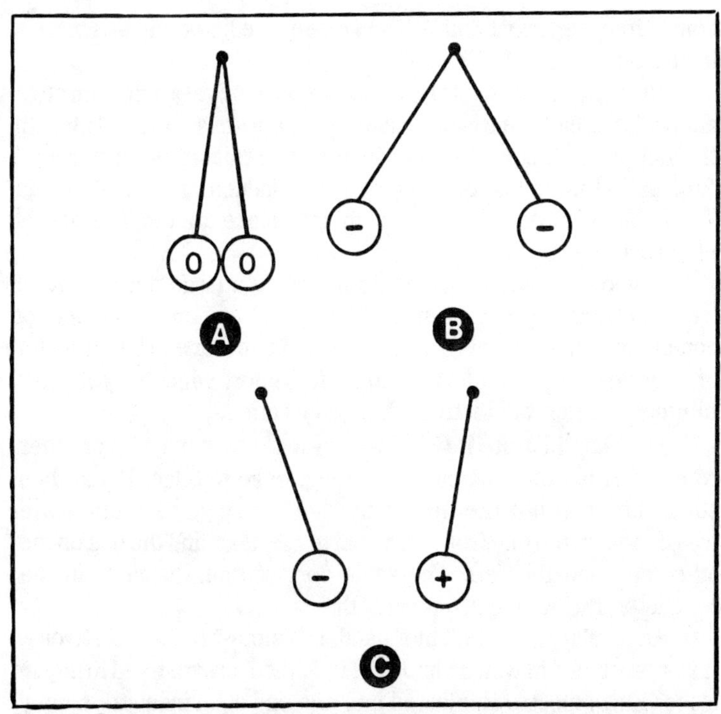

Fig. 2-2. One of the earliest electrical observations, which forms a key point to present-day circuit theory, was that objects with similar electrical charge repel each other, while those with opposite charge attract. Ping-pong balls hung by threads from a support can demonstrate the effect. With no charge (A) two balls hang side by side. When both are given negative charge (B), they fly apart from each other. If two are suspended separately and given opposite charges (C), they attract—but if they touch, their charges will either neutralize each other and result in condition A, or one charge will overpower the other leaving both with the same charge (condition B).

In general, all materials may be divided into two major categories called "conductors" and "insulators," depending upon the percentage of free electrons in the material. The dividing line between conductors and insulators is fuzzy, however—a conductor is a material which has a large percentage of free electrons, while an insulator has almost none. Many materials, however, fall somewhere between "conductor" and "insulator" in the action of their electrons, and so we have two additional groups separating conductors and insulators. These are "resistors" and "semiconductors."

A resistor, or resistance material, has enough free electrons to permit electrical action—but they are bound to their corresponding atomic nuclei tightly enough to make it necessary to use considerable energy to make them move. We'll get into this more a little later.

Right now, we're interested in the way electrons flow through an available path, and what makes the path "available." Any path through a conductor, semiconductor, or resistor is "available." Any path through a conductor, semiconductor, or resistor is "available" for equalization of an imbalance in the number of electrons at its two ends.

A good conductor is a material having a large percentage of free electrons. Injecting a single electron at one end of a length of conductor will force a single electron out the other end by a transfer of energies from one electron to its neighbor, which may involve millions of different electrons during the action.

You can illustrate it (Fig. 2-3) with a row of pennies or other coins lying flat on a tabletop, touching, edge to edge. If you then slide one coin into the end of the line sharply, its motion will transfer down the line from each coin to the next, and the one on the other end will fly free. The same sort of thing happens in the conductor; the *free* electrons carry the energy.

An insulator, on the other hand, has almost no free electrons. It is a substance in which almost all the electrons are fixed firmly in place. Extra electrons injected into one end of an insulator simply remain there, unable to transfer their energy to anything else, until a conductive path is supplied to permit them to flow.

A resistor has some free electrons but not as many as a conductor (the dividing line is virtually invisible; all known conductors have some resistance—as do all known insulators). In a resistance, some of the extra electrons injected strike atoms and cause them to vibrate. This vibration is what we know as "heat," and the result is that the energy of the extra electrons is converted to heat, or "dissipated." This removes electrical energy while producing new heat energy; the total energy remains constant but its form changes.

Any flow of electrons or electrical charge is known as electric current. The name dates back to the days when electricity was believed to be a "fluid" of some sort—complete with currents, waves, flow, pressure, and the like. Most of these words are today part of the electrical language—but with meanings that have little to do with fluids or gases.

TWO KINDS OF ELECTRICITY

A bit earlier, we mentioned that it was possible to "charge" an object by friction, which literally rubs off some electrons and thus

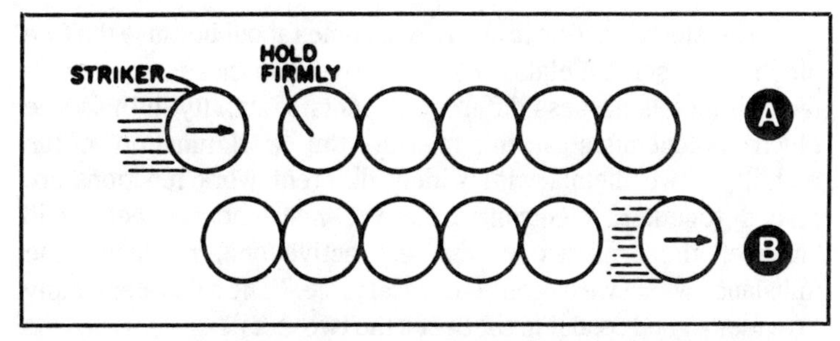

Fig. 2-3. The experiment shown here demonstrates how electric current is carried through a conductor by free electrons, without the individual electrons having to move all the way through the conductor. At top, a row of coils is placed flat on a table, edge to edge, and one coin (left) strikes the end of the row. When it hits, bottom, the impact will travel through each coin in the row, and the one at the far end will fly free. The popular swinging-ball toys sold as "executive pacifiers" also demonstrate the same principle.

temporarily destroys the inherent balance present in almost all natural materials.

This kind of electrical condition is known as "static electricity." The word "static" comes from the same root as does "stationary," and means the same—not moving. The ancient Greeks knew about static electricity; until the late 1700s no other kind was known.

Static electricity produces many interesting effects, but it cannot do any useful work. Only electrons in *motion* can do any work, and the only motion of electrons involved with static electricity occurs in the instant of discharge.

This not to say that static electricity can be ignored. Lightning, for example, is the discharge of static electricity built up in clouds and other air masses. We must know how to handle, and how to protect ourselves against, static electricity in order to make use of radio. For our useful work, though, we depend upon another kind of electrical effect—that produced by electrons in motion, and originally called "dynamic" electricity.

Although the buildup of a static charge by friction was the only method known to generate electric effects for many years, it is not the only method. Modern electrical science dates rather precisely from the discovery, in 1790, that electrical effects can be generated by chemical means. Today, we call such a device a "battery" or "cell"; originally, it was called a "voltaic pile." The first voltaic piles (Fig. 2-4) consisted of stacks of different kinds of metals (silver and zinc were favored) separated by blotting paper soaked in saltwater.

The electric action in such a pile comes about because the free electrons in some metals are more active than those in others. In fact, all metals possess different amounts of activity in their free electrons (chemists call this measure the "work function" of the metal). If two metals with widely different work functions are placed together, electrons from the more active metal will "invade" the structure of the less active one, producing the imbalance which we call an "electrical force." The saltwater simply provides a good conductor between the two metals.

Since the force in this case is caused by chemical differences between the metals or "electrodes," it is a continuous rather than a temporary state.

This is, incidentally, the reason why boat owners must watch out for "electrolysis" of the metal parts of their vessels. The boat actually becomes a battery when electrolysis is present, and the more active of the two metals is literally eaten away by the process.

Once a steady source of electricity was available in the voltaic pile, experimentation boomed—and other methods of "generating" electricity were discovered. The one most widely used today is the electric dynamo, which moves a magnetic field past a conductor to "induce" a current in the conductor. To explain how this happens it would be necessary to go far more deeply into physics than is necessary for the Novice license. It is, however, the way ordinary house current is "made."

CURRENT—STEADY AND UNSTEADY

At this point, we have seen that a flow of electric charge is known as an electric current, and we have met several methods for getting such a flow going. The methods include discharge of a static charge, chemical action in a battery, and mechanical action in a dynamo or generator.

All of the electrical effects which we make use of in radio depend in one way or another upon the flow of current, so it makes sense to look just a little deeper into what a "current flow" amounts to.

Current, in general, can be thought of as "electrons in motion," but we have to keep in mind one fact which is usually confusing at first. The names "positive" and "negative" were given to the two polarities of electric charge way back there by Ben Franklin; he had one chance in two of being right—and he lost. By the definition he proposed, which had come into common use

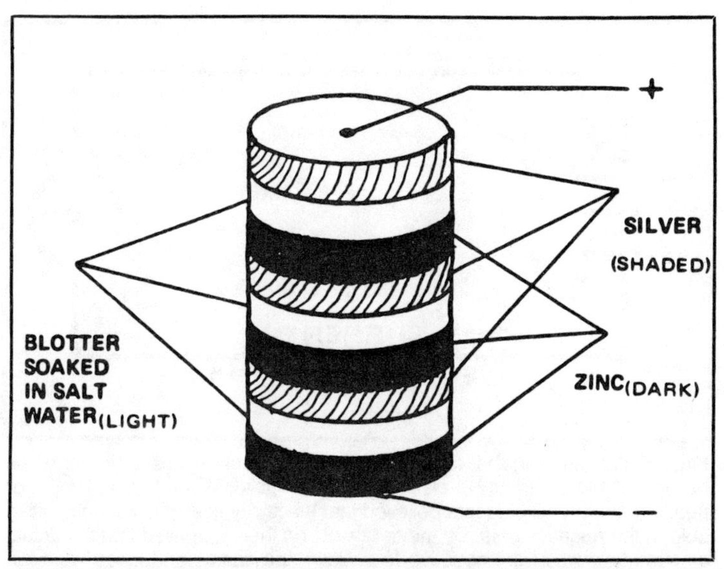

Fig. 2-4. First source of steady current available to electrical experimenters was the "voltaic cell" shown here. It consisted of a stack of alternating zinc, blotter, and silver disks, with the blotters soaked in saltwater. You can make one using silver coins and the newer zinc—copper ones, or even substituting copper pennies for the zinc. A silver—copper stack will develop about 0.5 volt per cell; the stack shown here contains three cells (one cell is one set of zinc, blotter, and silver plates).

before anyone knew better, "current" flows from positive to negative.

Electrons, however, are negative charges, not positive. They flow away from the negative pole of a battery or generator, not toward it. The result is that *current* flow, and the *electron* flow which it represents, are going in opposite directions (Fig. 2-5). A small point, perhaps, but it has confused many. If it troubles you, just think in terms of *electron* flow and don't worry about the direction of the current. Old Ben's mistake is too deeply embedded in the heart of physics to do anything about correcting it now!

Current is measured in units called "amperes," which are defined directly in terms of the amount of charge which passes a given point during a specific time. One ampere is the current which will carry one coulomb of charge past the measuring point in one second. The coulomb is defined as 6.28×10^{18} electrons worth of charge, but that's not very essential, because an alternate definition of the ampere is made in terms of its magnetic effect—and that's the one normally used in calibration laboratories.

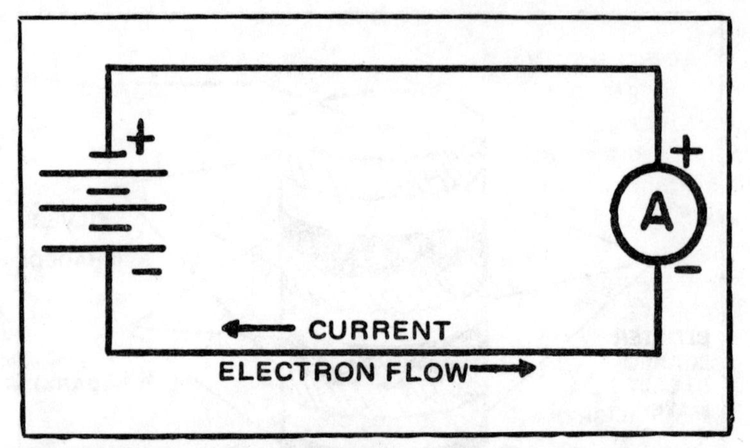

Fig. 2-5. With an even chance of being right, Ben Franklin guessed wrong when he named the poles of his electrical circuits "positive" and "negative." For nearly a century all students believed that electricity flowed from the positive pole to the negative pole. By the time electron theory showed that the actual electron flow was from negative to positive, the names and ideas were too deeply embedded in physics to be able to correct them. The result is the somewhat confusing situation shown here. "Current flow" is still from positive to negative, but it's actually the result of "electron flow" which moves from negative to positive. If you find this a bit too much to stomach, don't worry about "current"—you can handle all the theory easily by keeping track of "electron flow" only.

For our practical purposes, we can use still a third definition of the ampere: that current which a 1-volt force will push through a 1-ohm resistance. More on this relation will appear a little later.

A key point to remember about current is that *time* is an essential part of its definition. Many of us tend to think that current means the *amount* of charge; few of us remember that it's actually the amount *per unit of time*, but that makes all the difference in the world. It's like the difference between distance (in, say, miles) and speed (in miles per hour). Unless the charge is moving, current cannot exist.

And, conversely, whenever electric charge is moving, current flows. It's possible to move a quantity of charge a microscopic distance in one direction, then reverse it and move it back to its original position, repeating this action continually so that the actual average movement of the charge is zero—but a current flows.

Normal household electricity, in fact, works just this way. Such a current is called "alternating" current, or ac, because it alternates its direction of motion. The steady current from a battery, which always goes the same direction, is known as "direct" current, or dc (Fig. 2-6). Ac behaves differently from dc, and at the right time we'll examine its action. At this stage, however, we're still working with the most basic points, and we can bring them out better if we ignore ac for a while and concentrate upon dc actions.

The flow of dc depends upon three major factors. They are the inherent capability of the source (how many electrons it can supply), the "pressure" or voltage at which the electrons are supplied by the source, and the resistance of the circuit through which the current flows.

VOLTAGE, OR ELECTROMOTIVE FORCE

We just mentioned that one of the three major factors affecting the flow of direct current is the "pressure" or voltage at which electrons are supplied by the source. What is meant by "voltage"?

One of the original terms for the factor we now call "voltage" was "electromotive force," which is abbreviated emf and which may appear in the FCC examinations since it is still in use among physicists. It's essentially the same force which makes like charges repel each other.

Returning to static electricity for a moment, remember that one of the most common methods of generating a "static charge" is to rub electrons physically from one material to another by friction. Since each electron has a definite electric charge, the amount of charge which would result depends upon the number of electrons which move out of "balance" positions. That is, the amount of imbalance is a measure of the amount of charge. The greater the imbalance, the stronger the force which tends to accomplish rebalancing.

A lightning bolt (Fig. 2-7) is a good example of this. Normally, air is one of the best insulators known. When the electric force exceeds a certain high value (about 3000 volts per inch for dry air)

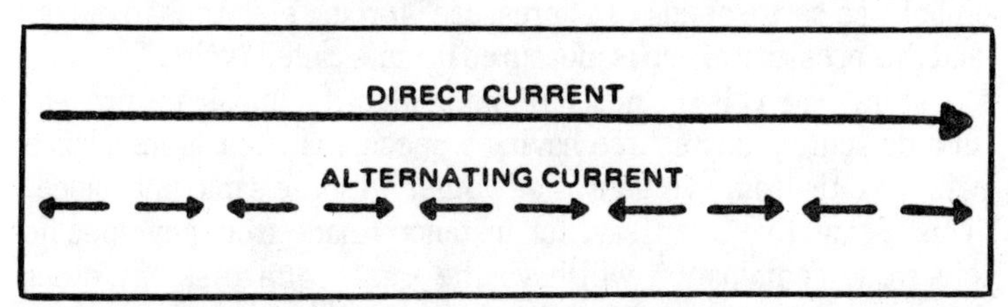

Fig. 2-6. Difference between "direct" and "alternating" currents is illustrated here. Direct current, top, flows steadily in same direction at all times. Alternating current, bottom, flows in short bursts, with direction reversing at end of each burst. Each can develop same amount of power, but the two types of current behave very differently. Radio waves are generated by ac.

Fig. 2-7. Lightning bolt illustrates pressure effect of voltage. Friction of dust in air, as well as ice crystals in cloud, charges thunderclouds continually. Because of electrical attraction, equal positive charge follows cloud across ground as it moves. When charge builds up pressure or voltage stronger than intervening air can take, air breaks down and forms a conductive path. Cloud then discharges, and woe unto any objects in that path since power levels are in the billions of watts.

though, the air molecules break down into "ions" which are good conductors. A charge will build up on a cloud until the force exceeds the breakdown voltage of air, and will then discharge as a bolt of lightning.

In a direct-current circuit, just as in a static-electricity generator, the necessary imbalance exists. The strength of the imbalance corresponds to a "pressure" forcing electrons to move, and this pressure of emf is measured in units called "volts."

Since the volt is a measure of the electron imbalance present in a dc source, any source having a specific amount of imbalance will have the same voltage as any other with the same imbalance. This means that a battery, for instance, made from any specific electrode combination will have the same voltage as any other battery using the same kinds of electrodes, regardless of its physical size.

For instance, the common flashlight cell uses carbon and zinc as its electrodes, and develops an emf of about 1.5 volts. Any cell using carbon and zinc will develop this same voltage; that is why a huge ignition-booster cell and a tiny penlite cell produce the same voltage despite their differences in size (Fig. 2-8).

A lead—acid storage cell, on the other hand, develops 2.2 volts per cell. A 12-volt auto battery consists of 6 cells, with a normal rating of 13.2 volts.

These voltage ratings are inherent in the chemistry of the substances used as electrodes. Different materials produce cells having different voltages.

Voltage is comparable to the idea of "pressure" as applied to liquids. The greater the voltage, the more current will flow (provided that circuit resistance remains unchanged). If 1 volt will force 1 ampere of current through a circuit, 2 volts will produce a current flow of 2 amperes, and 10 volts will result in 10 amperes of current.

The amount of *work* which can be done by electrical energy depends upon both the voltage and the current. This is just another way of saying that the work depends upon the absolute amount of energy in the circuit, because current measures energy per unit time. If the voltage is increased and current goes up, more energy flows in the same amount of time. Work or "power" is measured in watts, which are difined as "volts times amperes."

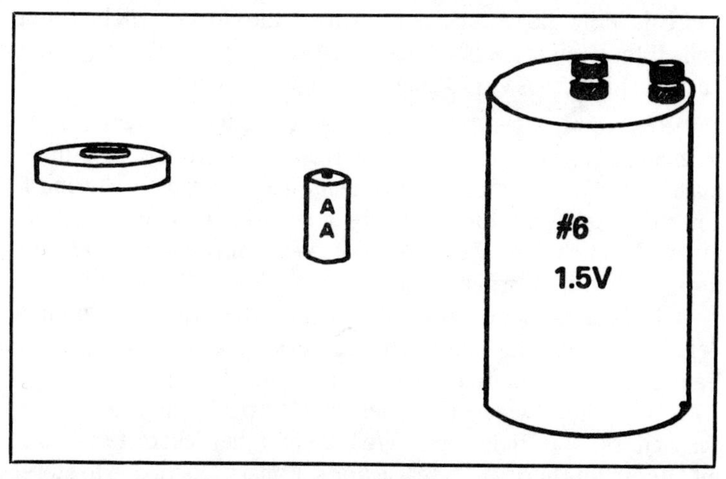

Fig. 2-8. Physical size of battery has nothing to do with its voltage; only factor setting voltage is material from which cell is made (and number of cells in battery; all these are single cells). Smallest AA size penlite cell and largest ignition-booster unit both produce 1.5 volts, since both use carbon—zinc electrodes. Larger cells will produce more current, or last longer at same current drain.

RESISTANCE, CAPACITANCE, AND INDUCTANCE

Another of the three factors which limits the amount of current flowing in a dc circuit was listed as "the resistance of the circuit." Actually, resistance is only a part of the picture—although with dc, it's the important part.

When we looked at the means by which current flows, and examined the differences between conductors and insulators, we found that many substances are neither good conductors nor good insulators. Such substances have enough free electrons to permit some current flow, but not enough to make it easy, and these materials are called "resistors."

"Resistance," then, is the capability of converting electrical energy to heat energy, as happens in these "resistors." We carry the idea a step further, and label *any* effect which causes electrical energy to apparently disappear as a "resistance." A transmitting antenna, for example, radiates the electrical energy applied to it—and we say that it does so because of its "radiation resistance." In this case the energy is not converted into heat—but it does disappear from the immediate circuit, and it's easiest to account for it if we consider that it was gobbled up by a hungry resistance.

Resistance in a dc circuit is a direct limiting factor on the amount of current which can flow. If the voltage remains constant in a circuit while the resistance is doubled, the current will be cut in half. If the resistance is reduced to 10% of its original value, the current will increase to 10 times the original amount.

Resistors (concentrated chunks of resistance, not necessarily the materials between conductors and insulators) are one of the three major types of electrical components used in radio. The most common resistors (Fig. 2-9) are composed of a composition material which furnishes the resistance, encased in a plastic insulating shell, with pigtail leads.

Resistance is measured in "ohms;" the practical working definition for an ohm is that amount of resistance which will permit 1 ampere to flow when an emf of 1 volt is applied.

The other two major types of electrical components are "capacitors" and "inductors." We'll go into their characteristics a bit later in this series. The capacitor's primary function is to store electric charge (all electrostatic generators include capacitors to accumulate the static charge), while the main job of the inductor is to "store" current by converting it into a magnetic field. In dc circuits, a capacitor amounts to an open circuit and an inductor

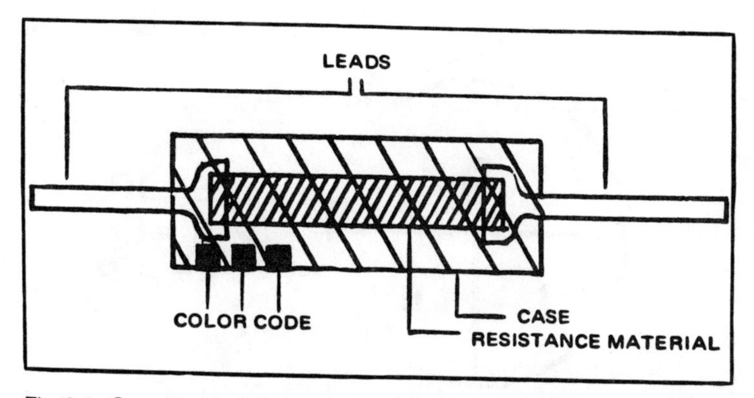

Fig. 2-9. Cross section of typical composition resistor (the most common kind) shows its construction. Pigtail leads permit it to be connected into circuit. Actual resistance material is inside plastic insulating case. Colored stripes around one end of resistor indicate value by means of standardized "color code," in which colors stand for numerals. Red, red, and brown stripes (reading from end toward center) would indicate 220-ohm resistor; red, red, and red would indicate 2200 ohms since innermost strips tells number of zeros to put on end of number.

behaves as a dead short; these components find their chief uses in ac circuits.

OHM'S LAW—THE FOUNDATION OF ELECTRICITY

By now you may have noticed that our definitions for the ampere, the volt, and the ohm tend to form a ring-around-the-rosy. One ampere is the current that flows through 1 ohm when 1 volt is applied. One volt is the emf which will force 1 ampere through 1 ohm of resistance. And 1 ohm is the resistance which will permit 1 ampere to flow when applied emf is 1 volt. And so forth.

These circular definitions are the actual *working* definitions we live with in radio, and they really are just as interdependent as they sound. What makes the whole thing hold together is our faith that at least two of the quantities we're interested in are known by other means at any one time.

Actually, these "definitions" are all expressions of the most basic rule in electricity. The rule was discovered in the early 1800s by a German schoolteacher named Georg Simon Ohm, and is known as "Ohm's Law" in his honor (the unit of resistance is also named for him).

At its simplest, Ohm's law states that the current flow in an electrical circuit is determined entirely by the voltage and resistance present, and current in amperes always equals voltage divided by resistance.

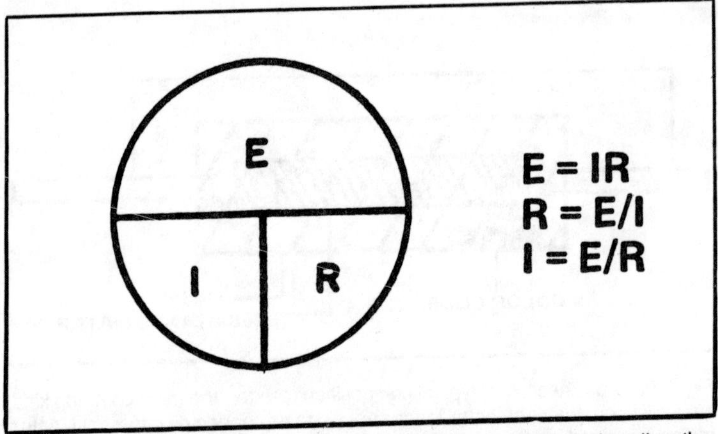

Fig. 2-10. Ohm's law, the cornerstone of all electrical calculations, describes the relationship between voltage, current, and resistance. Voltage is abbreviated "E" for "electromotive force" and current is abbreviated "I" for "intensity." The circle diagram may help you remember the relationship; by covering the symbol for the item you need to figure, the other two symbols appear in a format which indicates whether to multiply them or divide.

The law is usually seen in algebraic form, using the symbols "E" for voltage (electromotive force), "I" for current (intensity), and "R" for resistance. In this form, it becomes: E=IR.

This version is the one we used to define a volt. The law also appears in two other forms, to define current and resistance. The form which defines current is: I=E/R, and that which defines resistance is R=E/I (Fig. 2-10).

Using Ohm's law, we can find the current in any circuit if we know the voltage and resistance. Similarly, we can determine resistance if voltage and current are known, or voltage if we know current and resistance.

Ohm's law is used to calculate the resistor sizes necessary to produce a required voltage drop, since in this case voltage and current are known.

Because of its many uses and its fundamental nature, Ohm's law is the subject of questions on every class of amateur license exam above novice. The questions usually require you to *use* the law rather than merely quoting it, so it's a good idea to work problems with each of the three versions until you feel confident that you know which to use in any event and how to use it.

For instance, if you have a receiver which operates from a 6-volt battery and draws 2 amperes, but the only battery available is a 12-volt unit, what value of resistor would be necessary to reduce the voltage to a safe level? This is an Ohm's law problem in resistance, with both voltage and current known. However, the voltage actually used in the problem is slightly hidden; all we want to do is *reduce* the available voltage, not get rid of it all, so we don't use 12 volts in the problem. Instead, we use the *surplus* voltage, or 12–6=6V. We must get rid of 6 volts at 2 amperes; R=E/I, so we need a 6/2 or 3-ohm resistor to do the job.

A simpler problem is this one: What current will flow through a 30-ohm resistor if it is connected to a 12-volt battery? Here we want current, so we use the "I=E/R" form, and find that the current is 12/30, or 0.4 ampere.

Finally, a battery of unknown voltage forces 0.02 ampere through a 1000-ohm resistor. What is its voltage? Here we want to find voltage, so we use "E=IR" and learn that the battery's emf is 0.02 times 1000, or 20 volts.

Our first sample problem above is drawn directly from the ballast resistor of a 12-volt auto ignition system, while the last is a direct description of the most common type of voltmeter. Ohm's law is all around us—it is truly the foundation of electricity. Find other examples of Ohm's law in action, and develop additional practice problems to work until it becomes second nature to you.

Chapter 3
The Rest Of The Basic Theory

In order to qualify for any class of amateur-radio license, a would-be ham must prove to the FCC that he knows at least enough of the basic theory of radio to assure that he is capable of operating his equipment within the limits set forth by FCC rules and regulations. (He must also, of course, prove that he knows the limits set out in the rules—but that's the subject of a later discussion.)

So far, in this license study course, we have described and introduced a part of the basic theory necessary for the Novice license. Now, it's time to complete our look at basic theory so that we can move on to learn how transmitters and receivers work.

The preceding chapter described the electron theory, which offers a basis for describing (if not for completely understanding) electricity, and went through the rules which govern operation of direct-current circuits. To wind up this basic theory, we'll now examine alternating-current circuits.

AC-THE CURRENT THAT SWITCHES

Alternating current (ac) differs from direct current (dc) in only one essential point—but that one point makes a world of difference. For one thing, it makes possible the action of the electrical transformer, without which our nationwide network of electrical power systems could not exist as we know it today. For another, possibly just as important to our own special interests, it makes possible the existence of radio communication.

The one major difference is that direct current always flows in the same direction, while alternating current reverses its direction of flow periodically. This was mentioned in the previous chapter, which also included a sketch illustrating the difference—but we didn't go into its effects.

Before we get into the major effects, we can point up some of the reasons for them by looking at minor ones. For instance, a voltmeter or ammeter designed to measure dc will not indicate ac. This comes about because such a meter actually measures the magnetic force which accompanies a current. The direction of this magentic force depends upon the direction in which the current is flowing. With dc, which always goes the same way, the magnetic force is steady and pulls the needle across the dial. With ac, however, the magnetic force pulls first one way and then the other, and pushes and pulls cancel each other. If the ac is of extremely low frequency (below 10 hertz or so), you can actually see the needle vibrating; at higher frequencies, the needle cannot follow the rapid reversals, and so holds steady at zero.

Even though a dc meter indicates zero voltage and zero current, the power is still there—and anyone who has ever been bitten by ordinary household current in a defective appliance can testify to that. The meters simply are not built to respond to ac.

The reversal of magnetic field which accompanies the periodic reversals of direction of ac flow has effects more useful than making dc meters read zero, and we'll get into them shortly. First, though, we need to look at the characteristics which ac exhibits.

Like dc, ac has voltage, current, and power values associated with it in any circuit. None of them, however, are quite the same as their dc equivalents. Ac voltage, for instance, is always in a state of change. At one instant, it may be zero; a fraction of a second later, it can be +28 volts; still later, it's zero again, and the next time we look it may be– 10V. How can we measure something that's always changing?

The way this problem was solved was to decide that ac *power* should be the same as dc *power*; that is, a 1000-watt heating element should produce the same amount of heat with 1000 watts of ac that it does with 1000 watts of dc. Power, also, should be figured the same way—whether working with ac or dc—as volts times amperes. Once these decisions were reached, all that was necessary to define ac volts and ac amperes was to find a way of producing figures which would give the right "power" results.

Although ac reverses periodically, it swings between two peak limits, one positive and the other negative. Except for the difference of signs, these peak values are identical. An ac signal with a positive peak voltage of +28 volts would also have a negative peak voltage of −28 volts. The same is true of current, and these peak voltage values cen easily be determined with an oscilloscope (Fig. 3-1). The measurement, then, can be made in terms of "peak" voltages or currents, and converted to "effective" values which work properly with the power formula simply by multiplying them by a "correction factor."

As it happens, the proper correction factor to achieve this result is 70.7%, or one-half the square root of 2. (There are valid mathematical reasons for this, too—but they require the use of integral calculus to demonstrate.) Most ac voltage and current readings or ratings are in terms of this "effective" voltage or current, which is also known as "rms" for "root mean square," the name of the mathematical technique which derives the correction factor. That is, "110-volt" household power is actually 110 volts rms, or 155.54 volts peak. The "peak-to-peak" voltage is twice the "peak" rating, or a little over 311 volts.

In addition to voltage, current, and power, ac has two more characteristics which dc does not have, which are directly connected with the periodic reversal of current direction. They are *frequency* and *wavelength*.

Frequency is simply a count of the number of times the current reverses in a given period of time (usually one second). Until

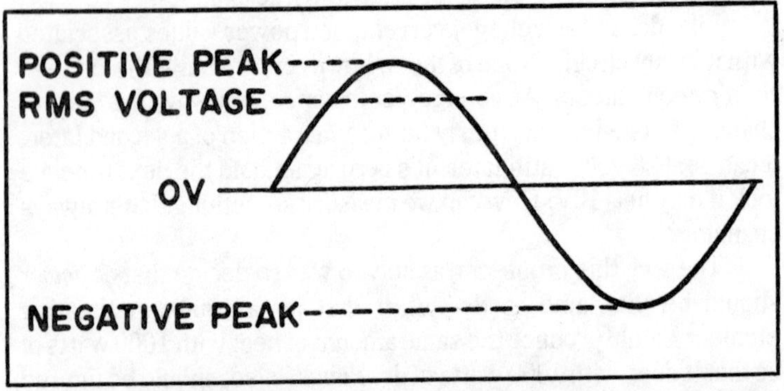

Fig. 3-1. Since both voltage and current values of ac signals are always changing, as shown here, it's not possible to determine exactly what the voltage or current is at any given instant. Because of this, ac voltages and currents are specified in a number of ways. One is by "peak" values, which are the greatest values reached at any time during a cycle. However, peak voltage times peak current gives power figures which are four times as large as they should be, so for most purposes "effective" (or rms) values, which are approximately 71% of peak values, are used.

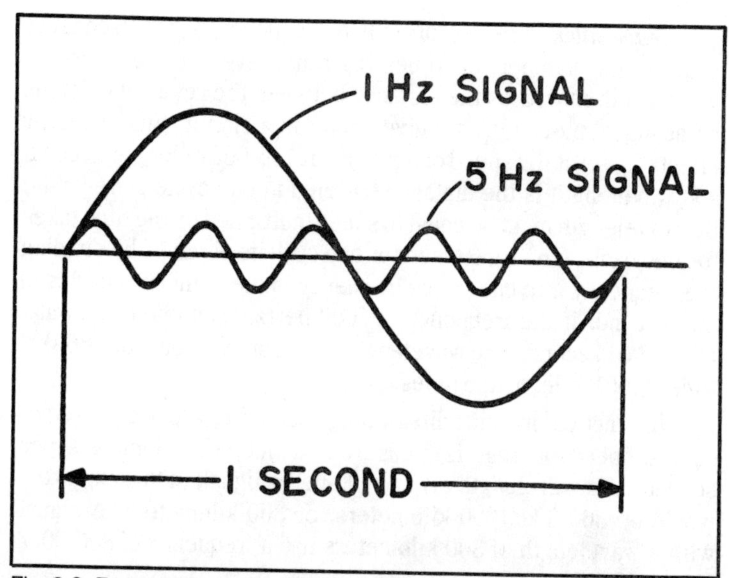

Fig. 3-2. Frequency and wavelength are two major characteristics of any ac signal. Frequency is the count of the number of complete cycles which occur during a given period of time, usually one second. Wavelength is the distance that the signal will travel during the time occupied by one cycle, at the speed of light. Frequency in hertz times wavelength in meters equals 300 million, since the speed of light (and radio signals) is 300 million meters per second. Since the 5 Hz signal's individual cycles in this sketch are of very short duration (1/5 second), they won't get as far during their "period" as the longer-duration signals—so their wavelength is smaller. But since more of these cycles occur in one second, their frequency is correspondingly higher.

recently, it was measured in "cycles per second" (cps or c/s) and some FCC questions still use this phrase. Receiver dials are sometimes marked in "kilocycles" which means merely "thousands of cycles" per second. In the past few years, however, the U.S. has adopted international standards which define the unit of frequency as the "hertz" in honor of Heinrich Hertz, the first man to demonstrate radio action. One hertz is one cycle per second. Thus, if an ac signal reverses direction 60 times every second, it has a frequency of 60 hertz; this is the frequency of standard household power. All ac signals are characterized by frequency (Fig. 3-2); some ac has frequency measured in tenths of hertz—one reversal in 10 seconds, and some is measured in millions of gigahertz-1,000,000,000,000,000 reversals per second. Radio signals range from 200 kHz up, with the major ham activity in the 3 through 30 mHz region and much VHF activity from 50 through 450MHz.

Wavelength is related directly to frequency. Where frequency counts the number of times current reverses per second, wavelength measures the distance between the reversal points in an ac signal moving past a given point in a circuit. Since electric effects travel at the speed of light, about 186,000 miles per second, and wavelength is the distance traveled in one cycle or reversal, the wavelength must be equal to speed multiplied by the time taken for one cycle. The time taken for one cycle is obtained by dividing the frequency into one, since frequency is the number of cycles in one second. If the frequency is 1000 hertz, then one cycle must take 0.001 second. The wavelength of this signal would be 186,000 times 0.001 mile, or 186 miles.

In practice, metric units rather than miles are used in radio, so the speed of light is measured as 300,000 kilometers per second. The wavelength of our 1000-hertz signal, in these figures, would be 300,000/1000 kilometers, or 300 kilometers. A signal with a wavelength of 300 kilometers has a frequency of 300,000/300, or 1000 hertz.

In the VHF region, wavelengths are much shorter; as a result, they are often measured in meters rather than kilometers. The same formula works, though, if you keep in mind that 300,000 kilometers is the same as 300 million meters. The wavelength of a 6-meter signal is 300 million/6 or 50 million hertz; we say for brevity that it's 50 megahertz, or 50 MHz.

Wavelength calculations are important when it comes time to erect a transmitting antenna, and often prove useful in installing feedline as well. More to the point, they are included in the FCC exams.

HARMONICS

In addition to the characteristics of voltage, current, power, frequency, and wavelength, ac signals have another quality not shared by dc—they have, and are, "harmonics."

A harmonic of an ac signal (Fig. 3-3) is another ac signal whose frequency is some exact multiple of that of the original signal. That is, a 120-Hz signal is the second harmonic of 60 Hz (2 × 60), the third harmonic of 40 Hz (3 × 40), the fourth harmonic of 30 Hz, and so forth.

Since the multiplying factor can be any whole number, every ac signal has an infinite number of possible harmonics. Usually, though, only the lowest ten or so are of any importance, because the higher the harmonic, the less likely it is to be generated in any detectable strength.

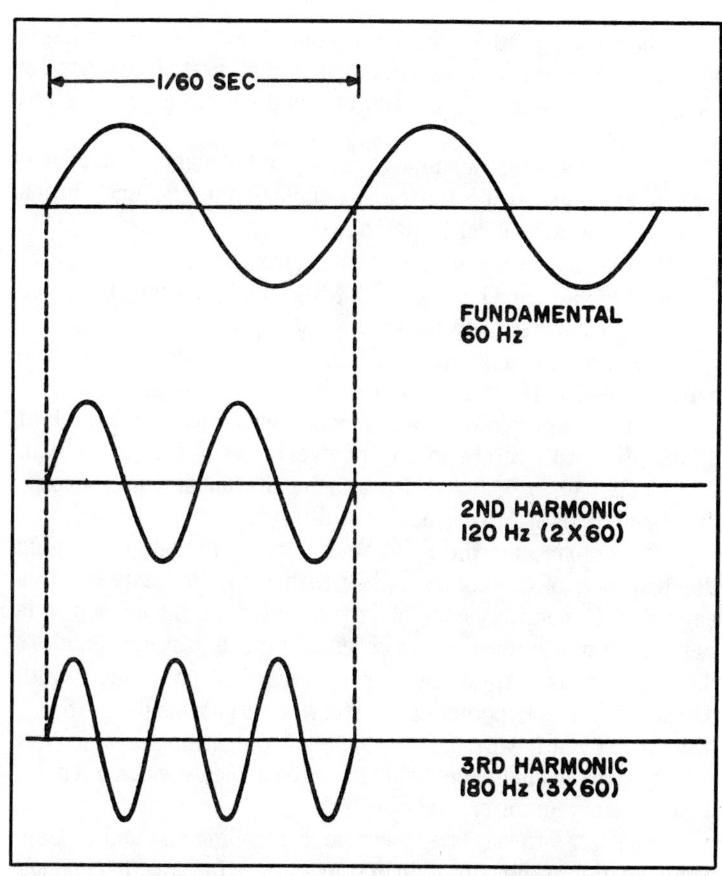

Fig. 3-3. Waveforms of three signals (fundamental and first two harmonics) are shown here to same scale. Note that each crosses "zero" line at same time as fundamental.

Similarly, any ac signal must be a harmonic of an infinite number of lower frequency signals; but again, only the "low order" signals have any practical effect in most cases.

The existence of harmonics is both a blessing and a curse to radio operators. The curse comes about because whenever an ac signal is present, its harmonics may easily be generated by accident. Some types of circuits generate harmonics as a normal part of their operation, while others generate them only if misadjusted. In any event, a radio transmitter is prone to radiate harmonics of its intended output signal, and careful operation is necessary to keep these unwanted output signals at harmonic frequencies from being radiated.

The blessing is that the ease of generating harmonics makes it simple to produce high-frequency signals of accurate frequency, by using low-frequency signal sources and creating the desired harmonics.

Way back in the beginning, far-sighted planners took advantage of this by getting the most popular ham frequency bands assigned on a harmonically related basis.

The lowest-frequency ham band, for instance, is at 1.8 MHz. The next higher band begins at 3.5 MHz, but the second harmonic of the first band is entirely inside this one.

Similarly, the next higher band runs from 7.0 to 7.3 MHz; the band above that is from 14.0 to 14.45 MHz, and so forth.

Thus, it's possible for a ham who owns only one crystal, at 3.505 MHz, to operate in any of five bands—at 3.505, 7.010, 14.020, 21.030, or 28.040 Mhz, by using the fundamental, second, fourth, sixth, or eighth harmonics of his crystal.

The emphasis in the FCC examinations, though, centers on the bad side of harmonics rather than on their benefits. This emphasis is natural, since if an unwanted harmonic signal is radiated, it may interfere with essential radio messages outside of the ham bands. While the second, fourth, sixth, and eighth harmonics of that hypothetical crystal were in ham bands, the third, fifth, and seventh were not—and even those harmonics which fell in ham bands did not necessarily fall into subbands which might be legal territory for the crystal's owner!

We'll get into ways and means of controlling harmonics later, when we look at the operation of transmitters for now, it's enough to know that the critters exist, and to know what they are.

TRANSFORMERS

A while back, we noted that ac will not show up on a dc meter because of the fact that the magnetic fields associated with its current reverse whenever current flow reverses, and we observed that this reversal had effects more useful than its cancellation of meter readings. Now we're ready to look at these more useful effects.

We know that a current has a magnetic field associated with it, and that a moving magnetic field "induces" a current in a fixed conductor. This is the basis of operation on the electric generator.

The magnetic field need not necessarily be in *physical* motion for this effect to occur; all that is required is that the strength of the field be changing, at the conductor's location. Because of this, any change in current in one conductor can "induce" current in another conductor adjacent to it, without any actual physical connection between the conductors.

If both conductors are coiled, the current-to-magnetic-field effect is intensified; the more turns in the coil, the greater the intensification. And if each conductor is coiled about a "core" of soft iron or special steel, the intensification will be even greater.

Such a device, consisting of two (or more) sets of windings on a common core (Fig. 3-4), is known as a "transformer." One of the windings is called the primary, and all others are secondaries. When current is applied to the primary, pulses of current are induced in each secondary.

With dc, unfortunately, as soon as the initial rush of current through the circuit passes and current flow reaches its "steady" state, induction of current in the secondaries comes to a halt because the magnetic field stops changing.

But with ac, the current and magnetic field are always reversing direction, and to do this they must change in strength. In fact, they're *always* changing in intensity—which means that

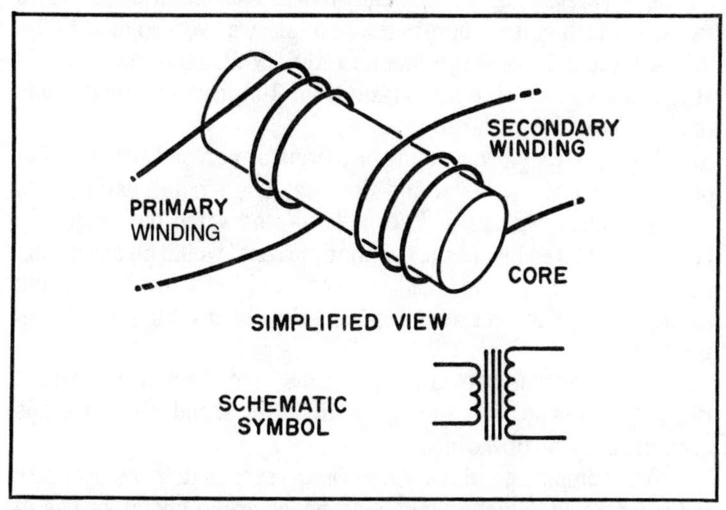

Fig. 3-4. Transformer depends upon conversion of energy from electrical to magnetic form, and back again. Simplified diagram at left is intended to show functioning; actual transformers use laminated cores of E-shaped sheet steel stampings, or donut-shaped "toroid" cores, rather than solid rod shown, and primary and secondary windings are usually atop each other rather than side by side. Current flow through primary produces magnetic field in core, which induces current in secondary as it changes. Schematic diagrams use symbol at right to indicate transformer. Inductors are similar, but are shown with only one winding.

corresponding alternating currents are induced in each secondary winding of the transformer.

If primary and secondary have the same number of turns, current in the secondary will be equal to that in the primary (less a small amount lost in transit; this is normally only a few percent, as the transformer is the most efficient energy-transfer device known).

If the numbers of turns on primary and secondary differ, the currents will also differ, because the relationship between current and magnetic field strength is constant and with fewer turns on one winding, more current is necessary in that winding to account for the energy in the magnetic field.

Thus, if the secondary has twice the number of turns that the primary has, it will have only half the current.

This property of changing current levels is what gave rise to the *transformer*; it transforms currents from one level to another.

But since power, also, is constant, and power is equal to voltage times current, when the current is cut in half the voltage must double. Therefore, the transformer with twice as many turns on the secondary as on its primary (a 2:1 turns ratio) produces twice as much voltage across the secondary as was applied to the primary (or a 2:1 voltage stepup ratio). Voltage stepup ratio is always the same as the turns ratio (Fig. 3-5); current stepdown is the inverse of the turns ratio.

If we had only one winding on a transformer, and attempted to push ac through it, we would find that this single winding was acting as both primary and secondary—and when the "induced" current was building up, the "applied current" would be increasing and vice versa. The induced current would actually be bucking against the applied current, opposing flow of any current at all in the circuit.

This quality is known as *inductance*, and is a valuable part of radio. It makes possible tuning circuits, filters, and other methods of controlling the flow of ac.

Any component which shows *inductance* as its main property is known as an inductor. Inductors which are intended for use in tuning circuits are commonly called *coils*, while those which are intended for filtering and for controlling the flow of ac are usually called *chokes*, for their ability to "choke off" ac. Chokes come in audio and rf sizes—and are often used in tuning circuits, since there is no clear-cut dividing line between a choke and any other inductor.

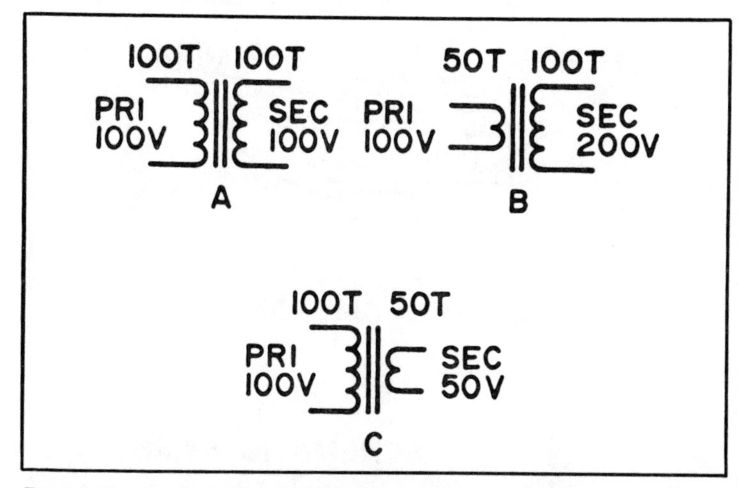

Fig. 3-5. Turns ratio (ratio between number of turns on secondary and number on primary) of transformer is important. With turns ratio of 1:1(A) output voltage and current from secondary are same as input values applied to primary. With 2:1 ratio(B), output voltage is twice the input voltage, and current is halved. With 1:2 ratio (C), output voltage is half the input value but current capacity is doubled.

INDUCTIVE REACTANCE

The choking action of an inductor is known as *inductive reactance* and is measured in ohms, like resistance. Reactance has many qualities in common with resistance, but differs in one major aspect—a reactance does not consume power. The opposition to current flow produced by a reactance comes about because energy is temporarily stored and then released out-of-phase with the applied energy, *not* because energy is converted from electrical form to heat.

CAPACITIVE REACTANCE

Inductive reactance is not the only kind; another kind, called *capacitive reactance*, also exists, and tuning circuits work by balancing one of these against the other so that they cancel each other and let current flow freely at some one single frequency, while opposing current flow at all other frequencies.

Capacitive reactance is a property of capacitors. A capacitor is in many respects like an inductor, but the resemblance is more theoretical than obvious at first glance. Any two conductors separated from each other by an insulator (Fig. 3-6) form a capacitor; most practical capacitors use very large conductors and

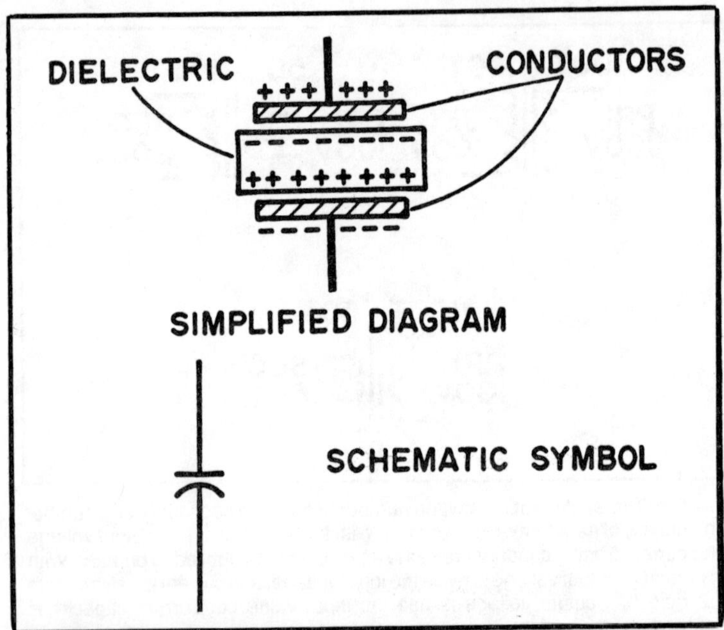

Fig. 3-6. Capacitor stores electric charge by physical displacement of electrons in the material of its dielectric (the insulating material separating the two conductive plates, at left above). When dc is applied, current flows until the charge is established and then stops. Since ac's direction of flow is always changing, charge can never be fully established and current appears to flow right through the dielectric. In schematic diagrams, capacitor is represented by symbol at right above. Curved plate in symbol represents grounded or lower-voltage side of capacitor if that is important, and is the outer plate in all cases.

very thin insulators or *dielectrics*, in order to get large capacitance in a small space.

When voltage is applied to such a circuit, dc cannot flow because of the insulator in the path. The pressure of the voltage, however, forces electrons from one side of the insulator to the other, and this amounts to an electrical charge within the insulator. The charge remains until it is either discharged by shorting, or leaks off through moist air.

Note that the capacitor works on voltage, while the inductor works on current. This difference is the major theoretical point which separated them; all the rest of the operation is almost identical despite the different appearance of coils and capacitors.

The capacitor blocks dc while allowing ac to pass through. The inductor does the opposite, passing dc and blocking ac. This makes possible filter circuits which can separate ac and dc which were originally on the same wire, and route each to a separate output lead.

Reactance also makes possible many other kinds of circuits, but a knowledge of them is not necessary for the Novice license.

We have now seen (at least to the degree of detail necessary for the Novice license) how ac and dc differ. It might come as a shock to some that two forms of electricity which are so different from each other could readily be converted, one to the other—but almost any radio makes use of this fact.

CHANGING AC TO DC

Ac can be changed to dc by simply putting an electrical "valve" in the circuit so that the current can flow in only one direction. Since the direction of current in this case cannot change, the current no longer "alternates" and must be direct (Fig. 3-7). It is, however, "pulsating dc," which changes in strength, and must be filtered to remove the "ripple" component and smooth it out into "pure" dc.

The most common such electrical valve is the "diode." Originally, diodes were a type of vacuum tube. Now, semiconductors are more often used, and the phrase "silicon diode" is almost a

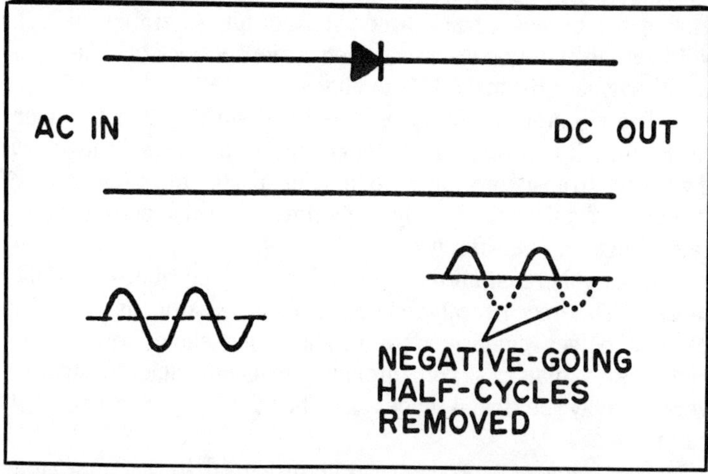

Fig. 3-7. Ac can be changed to dc by putting "electrical valve" in circuit to permit current flow in one direction while blocking it in the other. A diode is such a valve; output of this rectifier circuit is pulsating dc, and must be filtered in order to be useful. Rectification of ac is the standard method of providing dc operating voltage for all electronic equipment except for battery-operated portable gear. Circuits usually used are, however, a little more complex than the simple valve arrangement shown here.

single word in the radio vocabulary. The diode is, in fact, a valve for electrons, permitting them to move through it only in one direction.

FCC rules require that all transmitters operating at frequencies below 144 MHz be powered by "pure dc" to avoid excessive signal bandwidth. Most transmitters comply with this rule by converting ac from household power lines into the necessary pure dc with the aid of a transformer to adjust it to the required voltage levels, diodes to change from ac to pulsating dc, and filters composed of chokes and capacitors to remove the ripple and produce dc pure enough to meet the regulations.

CHANGING DC TO AC

Dc can be changed to ac by several methods. The "standard" method for many years was the "vibrator" (a buzzer-like device enclosed in a metal case and looking something like a vacuum tube), which interrupted the flow of dc through it. The output of the vibrator was often nothing more than interrupted, or pulsating, dc, but when it was passed through the primary of a transformer, ac came out of the secondary windings (Fig. 3-8).

This was the normal power supply for auto radios until development of low-voltage tubes, and later of semi-conductors. The 6-volt dc from the auto battery was converted to ac simply to enable the use of a transformer and permit generation of high voltage; the high-voltage ac was then converted back to dc just as if it had originated from the 110-volt lines.

The vibrator is still encountered occasionally, but has been made virtually obsolete by development of the power transistor. Two such transistors have no moving parts, they outlast the electromechanical vibrator by many times, and also generate less radio-frequency interference.

Ac is generated normally by either mechanical or electronic means. The mechanical generator, or alternator, is used by electric power companies to produce household power, and operates by moving a coil through a magnetic field. A smaller version may be found under the hood of any recent-model automobile.

The frequency of the current produced by an alternator depends upon the speed at which the alternator shaft is turning. The more rapid the rotation, the higher the frequency. Household power normally has a frequency of 60 Hz, while the ac output of an auto alternator ranges from 30 to 1000 Hz, depending upon engine speed.

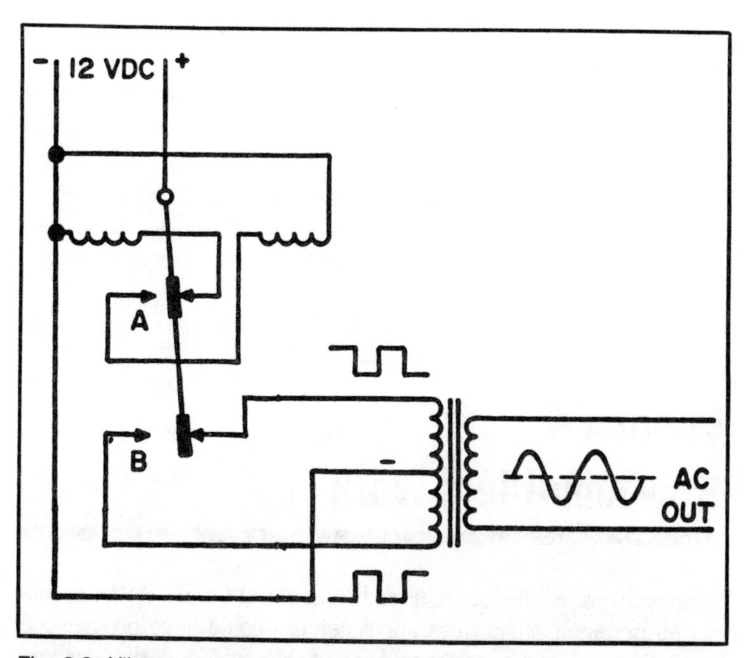

Fig. 3-8. Vibrator was standard means of converting dc to ac for mobile equipment for many years. Two coils in vibrator were connected, one at a time, to dc source by contacts at point A. Whenever one coil is connected, other is disconnected—and whichever coil is connected pulls the arm toward it, thus disconnecting itself and connecting the other one. This keeps arm vibrating, just like a doorbell or buzzer. Extra contacts at B switch dc between opposite ends of transformer primary winding, which reverses direction of magnetic field in core and produces ac in the secondary. Vibrator has now been largely replaced by transistor circuits which do the same thing without moving parts.

The electronic generator of ac is known as an "oscillator," and consists of an amplifier connected so that its output furnishes all necessary input, together with some "resonator" to determine the frequency. Oscillators can generate ac over a much wider frequency range than can alternators; normal range from oscillators is from less than 5 Hz up to frequencies in the hundreds of MHz, while alternators operate in the audio and low rf ranges only.

We'll take a more detailed look at oscillator action when we examine how transmitters work, since every radio transmitter contains an oscillator to generate the radio frequency involved.

You now have enough acquaintance with the basic theory involved in both ac and dc electronic and radio circuits to make your way through the Novice examination. In the next chapter, we'll apply some of this theory by taking a look at how amplifiers operate.

Chapter 4
How Amplifiers Work

Amplification is the heart of radio as we know it. Without this action neither transmitters nor receivers could have progressed much beyond the spark-gap days of Marconi and the earliest experiments.

And because amplification is so essential to radio today, the FCC rightfully requires all would-be hams to know at least a little about this action and how it works. The higher the license grade expected, the more detail is required—but even the Novice class licensee should have some knowledge of amplification and amplifiers.

The word "amplification" means "increasing the strength or amplitude of something," and in radio, this "something" is almost always a signal represented as an electrical voltage or current. We usually think in terms of voltages, but some types of amplifier circuits act on current instead of voltage, and in any event the "strength" which is increased is actually the *power* level of the signal rather than either its current or its voltage alone.

From this definition, it follows that an amplifier is something which amplifies, or in radio, is a device which acts to increase the strength or power of an electrical signal.

It's important to note, right here at the beginning, that in order for any device to qualify as an amplifier it must increase the *power* of a signal. A simple stepup transformer such as those we met in our previous chapter increases voltages, but it does not boost power and so cannot qualify as an amplifier.

Since an amplifier must increase the power of the signal applied to it, it follows that every amplifier must add energy from some outside power source (Fig. 4-1) to its input signal in order to produce an amplified output.

A number of devices are available to perform this function of adding power while being controlled by an input signal. The two most common (almost no others are currently used in radio) are the vacuum tube (Fig. 4-2) and the transistor (Fig. 4-3). Let's see how each of them, in turn, accomplishes its purpose.

HOW DO VACUUM TUBES WORK?

The simplest vacuum tubes used for amplification consist of three active elements inside a sealed enclosure. These elements are called the cathode, the grid, and the plate. Usually there's a fourth element, the heater, which serves only to keep the cathode hot enough for the tube to operate. The heater is electrically insulated from all the other elements, and takes no direct path in the amplifying action.

The tube is "filled with a vacuum"; that is, the space inside the enclosure or "envelope" contains as close to no atmosphere at all as we can possibly achieve.

Under these conditions, when the cathode is made sufficiently hot by heat energy from the heater, electrons at the cathode surface take on enough energy from the heat to fly off the surface and form an electron cloud or "space charge" surrounding the cathode as shown in Fig. 4-4.

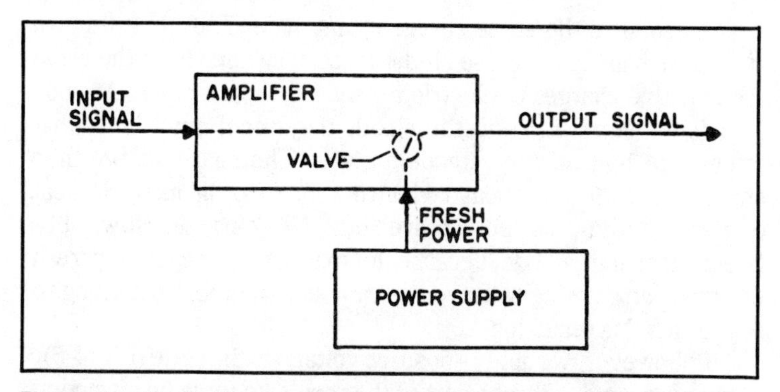

Fig. 4-1. Amplifier "adds power" to electrical signal. Only way this can be done is to use original signal to control strength of fresh power from some power supply as shown here; many types of "valves" are available to perform this control, but those most often used are vacuum tubes and transistors.

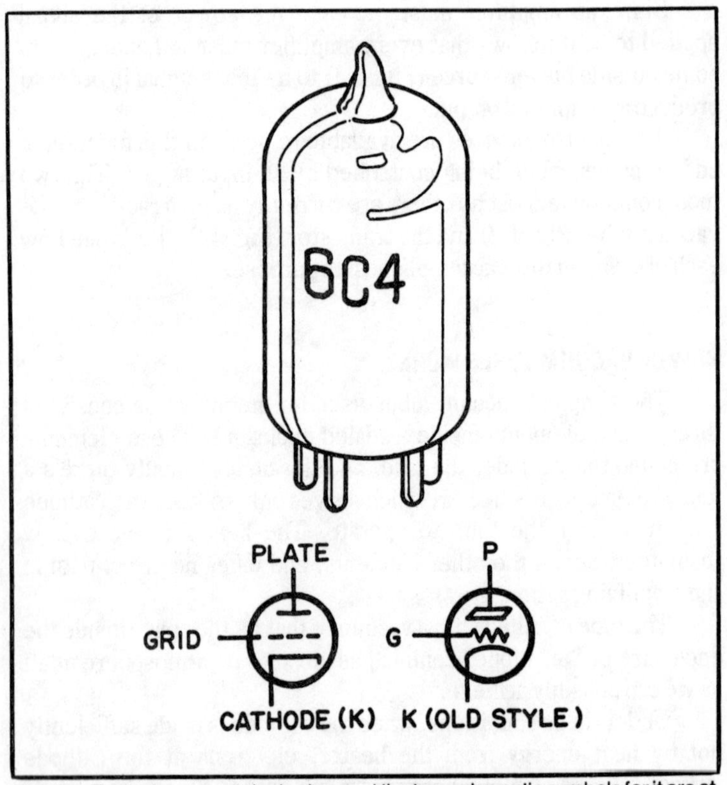

Fig. 4-2. Typical vacuum tube is shown at the top; schematic symbols for it are at bottom. This is a "triode" or three-element tube; several other types have different names, each with a prefix denoting its number of elements.

As soon as the space charge forms, it tends to limit its own size. This happens because all the electrons which form the cloud are negative charges of electricity, and repel each other. As soon as a few electrons are in the cloud, they repel any others that attempt to boil off the cathode surface. The space charge then stabilizes, with just enough electrons in it to balance the heat energy tending to boil more off. No current flows between plate and cathode, because the space charge repels any fresh electrons attempting to leave the cathode, and there's nothing to counteract this repulsion.

If, however, we apply a positive voltage to the plate (Fig. 4-5), it will overcome at least a part of the repulsive force by attracting some of the electrons in the space charge over to the plate. For each electron pulled out of the space charge, there's room for a fresh one to leave the cathode surface. The net result is that electrons keep moving from the cathode into the space charge, and from the space charge across the intervening empty space to the plate. They continue their journey through the external power source and back to the cathode, and so current flows in the external circuit.

Should the plate be made negative instead of positive, the electrons of the space charge would be repelled rather than attracted, and so no current could flow. This property makes the simple diode tube (two-element tube, containing only plate and cathode) effective as a "valve" for electricity. Current can flow through it only when the plate is positive.

While the diode is useful as a one-way valve for electricity, it cannot in itself amplify since it has no means of controlling the current flow. If any current can flow, it will flow at the maximum

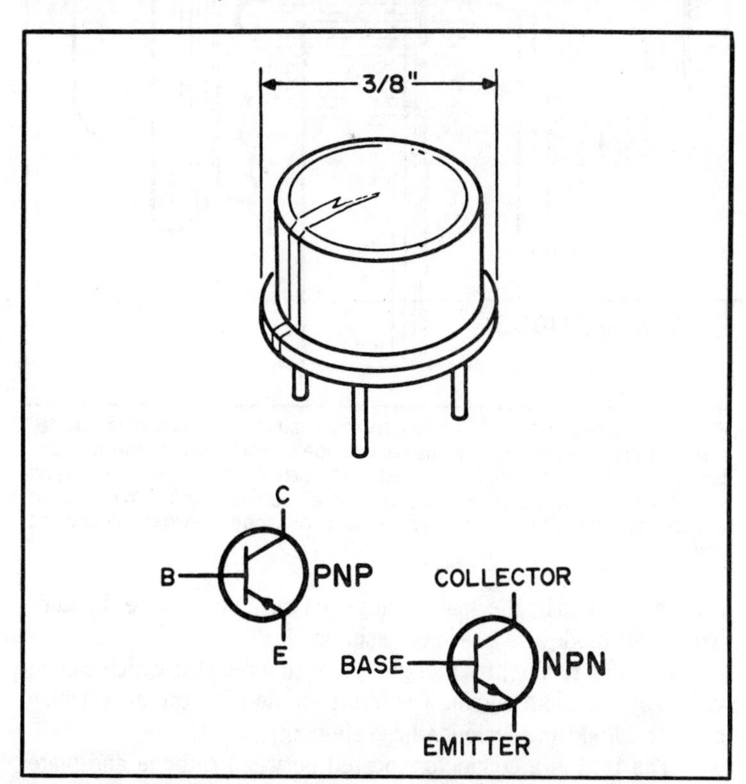

Fig. 4-3. Transistor does same types of jobs as vacuum tube, in much smaller space. Typical transistor is shown at left; this is "TO-5" configuration, and actual transistor element is much smaller (can provide protection against corrosion and so forth). Schematic symbols for both polarities of transistor are shown at bottom.

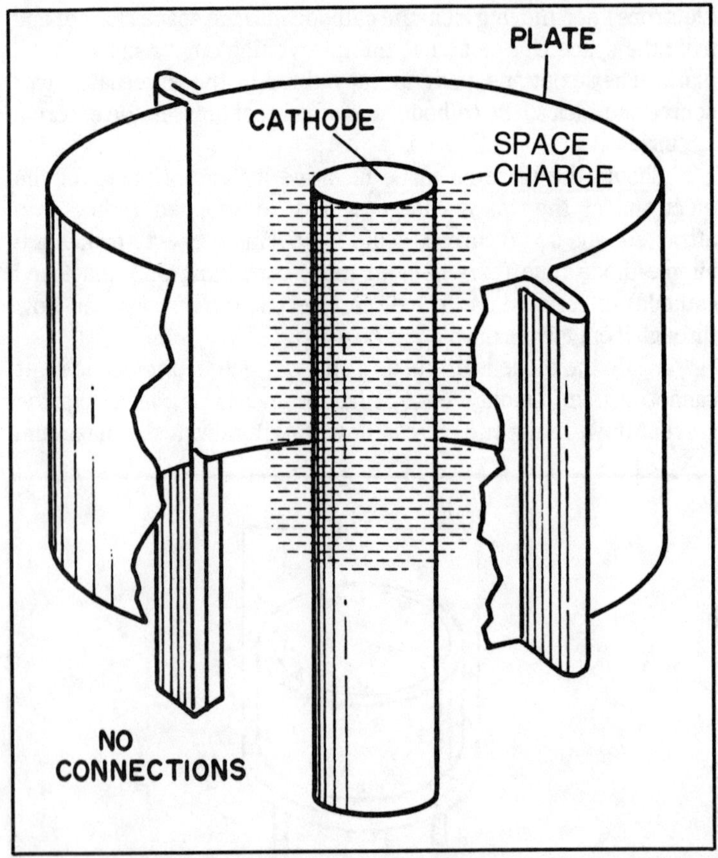

Fig. 4-4. Starting point for understanding vacuum tube action is "diode" (two-element tube) which contains hot cathode and cold plate inside vacuum. Electrons boil off surface of cathode, but repel each other once they are free and so form a tight cloud called "space charge" surrounding the cathode. In absence of external forces, no current flows across space between cathode and plate.

rate established by the space charge and the plate voltage. In other words, the diode acts as a fixed resistor.

In order to amplify, though, we need a resistor which can be varied by the input signal. That's the job done by the grid, which turns the diode into a triode (three elements).

The grid is a conductor located between cathode and plate (Fig. 4-6) so that the electrons streaming from cathode to plate must flow past or through the grid. Often the grid looks like an opened-out coil spring; sometimes it's built in a mesh form, like a screen for a window. At any rate, it contains openings through which most of the electrons leaving the cathode can flow to the plate.

Since the grid is between cathode and plate, it can exercise control over the electron stream. If, for instance, the grid is made negative with respect to the cathode, it will tend to repel the

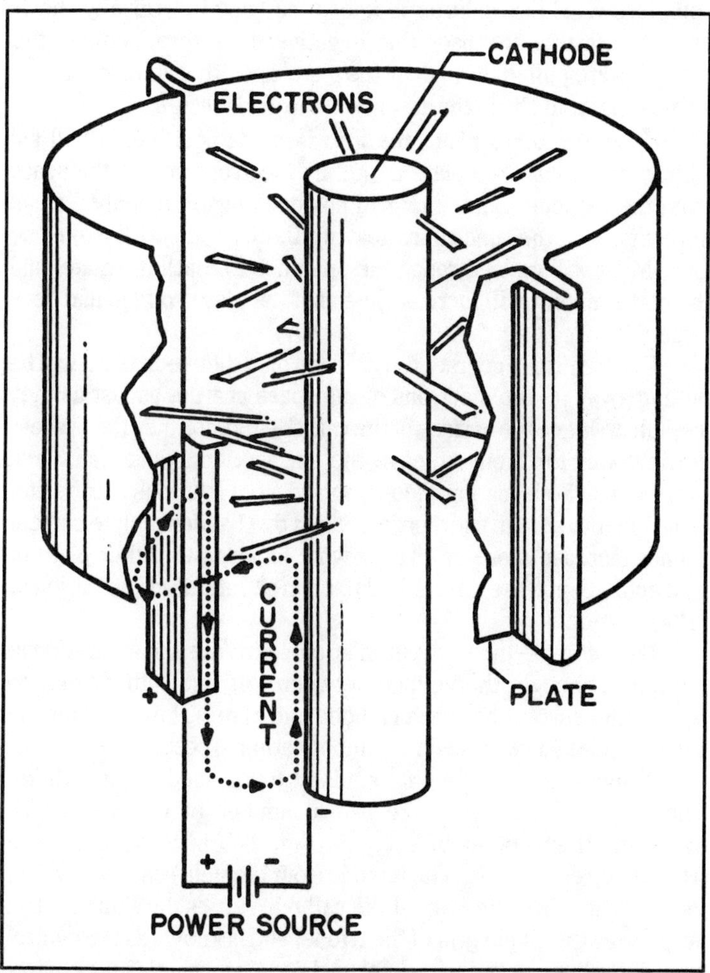

Fig. 4-5. When power source is connected to diode so that plate is made positive with respect to cathode, electric force attracts electrons out of space charge to the plate, and current flows across the empty space between plate and cathode. If plate were negative, electrons would be repelled instead of attracted. Thus diode is one-way valve for current; current flows only when plate is positive. This action is used in power supplies and for other purposes, but does not provide amplification since new power is not controlled by an input signal.

electrons. This repulsion by the grid tends to overcome the attraction provided by the positive plate voltage, because the grid is closer to the cathode than is the plate.

In fact, if the grid voltage is sufficiently negative, it will repel all the electrons of the space charge, and cut off plate current flow completely. This condition is known as "cutoff." Usually, however, the grid is not made that negative; therefore, some of the electrons are pulled through by the plate's positive voltage, but not so many as would be in the absence of the grid voltage.

If, on the other hand, the grid is made positive, it will aid rather than hinder the plate in pulling electrons across the space from the cathode. Since the grid has openings in it, only a small percentage of the electroncs leaving the cathode will strike the grid. Most will go on through the openings to reach the plate, and the plate current will increase just as if the plate voltage had been raised.

Eventually, of course, the grid and plate between them will be pulling away all the electrons of the space charge, as fast as they are able to leave the surface of the cathode. Sometimes the process doesn't wait for them to leave the surface, and yanks fair-sized chunks of cathode material loose; this is most injurious to tube life, reducing it to zero in short order. When this happens, plate current cannot increase more no matter what happens to either plate or grid voltage, and the tube is said to be in a "saturated" condition or in "saturation."

Between the limits of cutoff at one extreme and saturation at the other, though, the voltage between grid and cathode acts to control the current between cathode and plate. This provides the variable resistance we need for amplification to occur.

Triodes are not the only kinds of tubes used for amplifiers. The three-element structure has a number of limitations. To overcome them, other tube types (Fig. 4-7) have been devised which have extra grids. The tetrode (four element) uses two grids, for example. The grid nearest the cathode serves the same control purpose as the single grid of the triode, and is known as the control grid. The other, between control grid and plate, shields the control grid from the plate to reduce interaction between input and output circuits, and is called the screen grid (or simply the screen).

The presence of the screen grid makes the tube operate at radio frequencies with simpler circuits than are required by triodes, but it introduces another quirk. Whenever screen voltage happens to be greater than plate voltage (which can occur easily

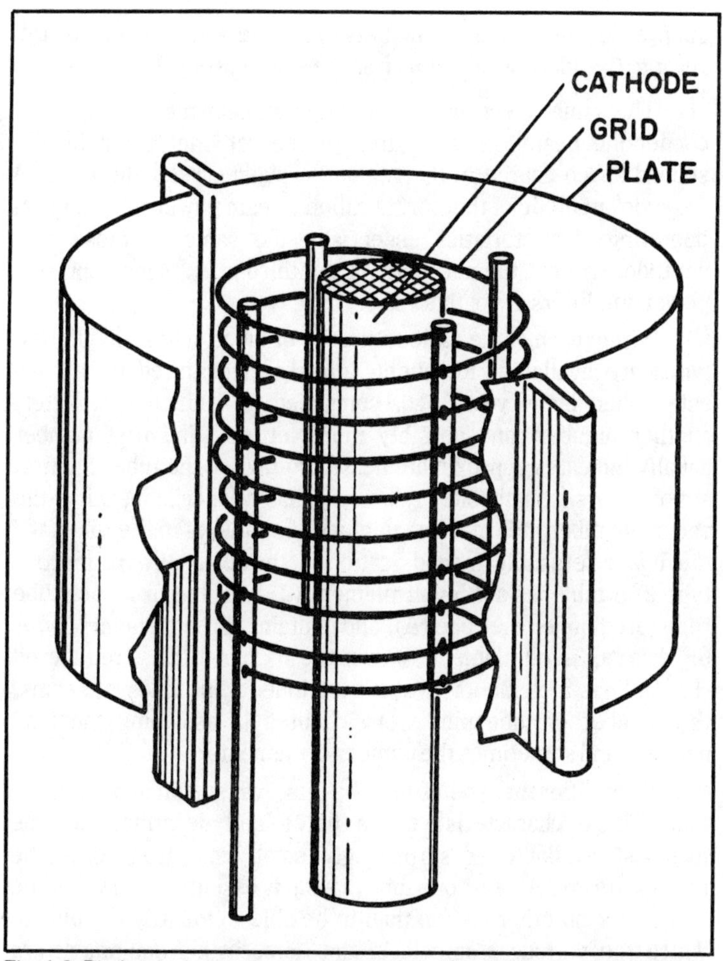

Fig. 4-6. Diode tube is excellent "one-wave valve" for electric current but offers no means of controlling current flow except on or off. Addition of "grid" between cathode and plate makes it possible to control the number of electrons which pass through it. Most grid structures are open at spiral of wire as shown here, but some types of tubes use a mesh like window screening.

during parts of the signal cycle), current may actually flow from plate to screen and cause undesired effects. The pentode (five elements) was developed to take care of this problem.

The pentode has three grids, one more than the tetrode. The added grid is located between the screen grid and the plate, and is connected either inside the tube or through an external jumper wire to the cathode. This puts a conductor between screen and plate which is more negative than either of them, and so suppresses any current flow between plate and screen. For this reason, the added grid is known as the suppressor grid.

The same effect may be obtained without the third grid, by connecting beam-forming plates to the cathode to establish a second space charge in the region between screen and plate. A "pseudo" pentode of this sort is called a beam power tube, and has operating characteristics essentially the same as those of a pentode, without the necessity of the third grid. Many popular rf power amplifiers are of this variety.

Within each of the various classes of tubes, a host of different types are available. Receiving tubes are identified by a code established many years ago, composed of a number, a letter, another number, and possibly more letters. The first number usually indicates approximate heater voltage of the tube; the first set of letters (usually one or two only) has no special meaning; the second number indicates the number of elements in the tube; and the final letters indicate modifications of the design. For instance, a type 2A3 tube has a 2.5-volt filament (2), was the first such tube registered under the code (A), and contains three elements (3). A 6F5 has a 6.3-volt filament (6), was the sixth of its class registered (F), and has 5 elements—yet it's a triode. Sometimes, it seems, the number of tube pins were counted in assigning the final number, and sometimes the number of elements!

Each different type of tube has its own specific characteristics. These characteristics are listed and described in tube manuals, available at surprisingly small cost from the tube manufacturers. At least one tube manual is essential for a would-be ham, if for no other reason than to be able to identify the pins to which the various elements of the tubes in his equipment are connected.

The characteristics of any tube are determined primarily by the physical construction of the tube. For instance, a large cathode structure with extensive surface area can release many more electrons at any given temperature than can a small one, and so a tube with a large cathode can carry a higher plate current.

The closer the grid is to the cathode, in comparison to the cathode-to-plate spacing, the more control the grid exercises over plate current. Consequently, a high-gain tube has its grid close to the cathode.

All of the tube characteristics are determined in this way, by the geometry of the tube's physical structure.

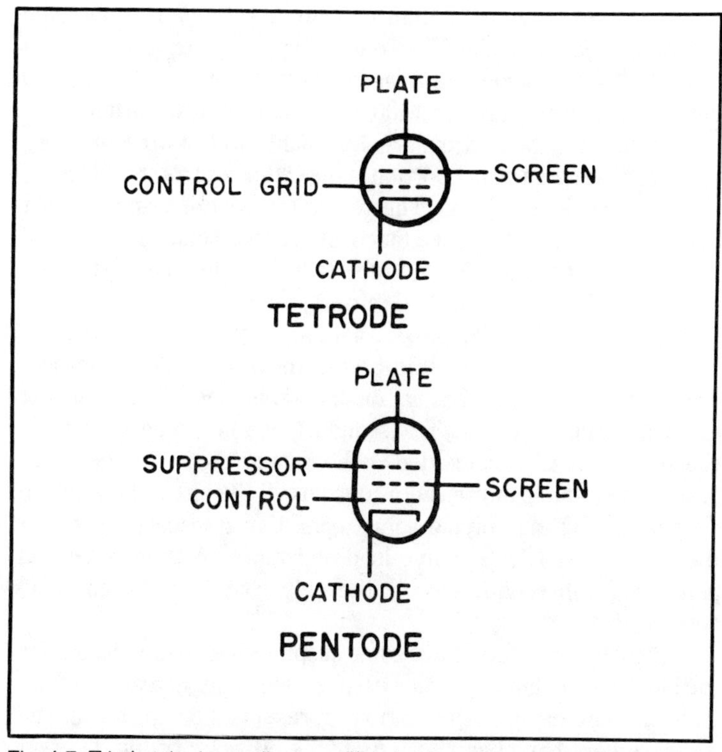

Fig. 4-7. Triode tube is excellent amplifier, but has some shortcomings which were overcome by adding additional grids to create "tetrode" and "pentode" types. Schematic symbols for tetrode are shown here.

Of all the characteristics possessed by a tube, three are of major importance to its users. These are the "amplification factor" which is often called its "mu" since the Greek letter mu (μ) is used by engineers to represent this ratio, the plate resistance, and the transconductance.

Amplification factor is a measure of the tube's gain. A tube with an amplification factor of 50, for instance, will show the same change of plate current with either a 1-volt change of grid voltage or a 50-volt change of plate voltage. Amplification factors for triodes range from 3 to 100 or so; "high-mu" tubes have factors greater than about 30, "medium-mu" tubes range from 8 to 30, and "low-mu" tubes run below 8.

While it might appear that the higher the amplification factor, the better the tube, that all depends on your purpose. In order to get high-mu performance, the grid-to-cathode spacing must be very small. This means that the cathode cannot be large, nor can it operate at very high temperatures, and therefore a high-mu tube cannot handle high plate currents. Typical plate current of a high-mu triode is about 1 milliampere (mA), compared to 10—25 mA for medium-mu tubes and 100 mA or more for low-mu units.

In addition, the grid must contain many turns of wire, or a very fine mesh, so that a larger portion of the cathode current is blocked, in order to achieve high mu. This means that a large change in plate voltage is necessary to make much difference in plate current, and so the plate resistance of the tube (which is the actual effective resistance of the plate—cathode circuit) is high.

Because of these factors, it's hard to judge the potential gain of a tube from either its amplification factor or its plate resistance. The transconductance rating, on the other hand, takes almost everything into account. Transconductance is a measure of the change in plate current divided by the change in grid voltage which produced it, and in any vacuum-tube circuit the circuit gain can be determined by multiplying actual operating transconductance of the tube by the true effective load resistance. With tetrodes and pentodes, transconductance is the only characteristic normally considered.

By now, you should have some suspicions as to the reasons for the hundreds of different types of tubes available. Just about every factor in tube design or operation represents a compromise, and the designers have been busy for some 50 years trying different ways of getting "the perfect tube." They're still trying.

HOW DO TRANSISTORS WORK?

Physically, the transistor looks like a completely different breed of device from a vacuum tube—and in many ways, it is. Where the tube is a glass or metal envelope containing at least three active elements plus a heater (for a simple triode), the transistor is a single crystal of solid material, with no heater, and just three connections made to the single slab.

However, the basic principles of the two different kinds of devices are surprisingly similar.

Just as the triode tube consists of a diode plus a third element (the grid) to control current flow, so is the transistor built up from diodes. In the transistor, however, they are "semiconductor" diodes rather than the kind used in the vacuum tube.

Semiconductors are materials which fall between those known as insulators and those known as conductors. The two most popular semiconductor materials, which account for just about all transistors available on the market, are germanium and silicon. Silicon is the major ingredient of common sand or glass; germanium, on the other hand, is comparatively rare.

Both silicon and germanium are chemical elements which form large single crystals, which will include within their crystal isolated atoms of other elements such as arsenic, phosphorus, antimony, boron, aluminum, gallium, or indium. These other elements are known as "impurities," and it's the impurities in a transistor or a semiconductor diode which make it work. They are, of course, carefully controlled—but they're vital to the operation.

Two different classes of impurities are used in semiconductors. One of them has one *more* electron per atom than the semiconductor material itself, and the other has one *less* electron in each atom. Arsenic, phosphorus, antimony, and boron all have extra electrons, while aluminum, gallium, and indium all are short an electron. Those with extra electrons are known as "donors" and those which are short an electron are called acceptors or receptors.

Semiconductor material which contains "donor" impurities thus has too many electrons to fit into its crystal structure, and would have a negative charge because of the surplus were it not for the fact that each excess electron's charge is balanced by the corresponding charge on the nucleus of its atom. Nevertheless, this type of material is called N-type (for negative).

That which contains "acceptors" impurities, similarly, would not have enough electrons to fill the crystal structure, and would have positive charge except for the balancing effect of the acceptor atomic nuclei. This type of material is called P-type (for positive).

A semiconductor diode consists of a single crystal of material, which is "grown" in such a manner that it is N-type at one end and P-type at the other (Fig. 4-8). The region where the two types meet is known as the junction of the diode.

At the junction, the excess electrons from the N-type material tend to cross over and fill the vacant spaces or "holes" of the P-type material. This pulls electrons away from their corresponding atoms, which upsets the internal balance of charges and produces a so-called depletion region around the junction which corresponds to the space charge of the vacuum tube. It's like having a built-in battery.

If we connect an external battery to this diode so that the P-type material is made more positive (Fig. 4-9), the negative voltage on the N-type material will push more electrons over into the depletion region. There, they will fill "holes" in the P-type material. At the same time, the positive side of the battery is

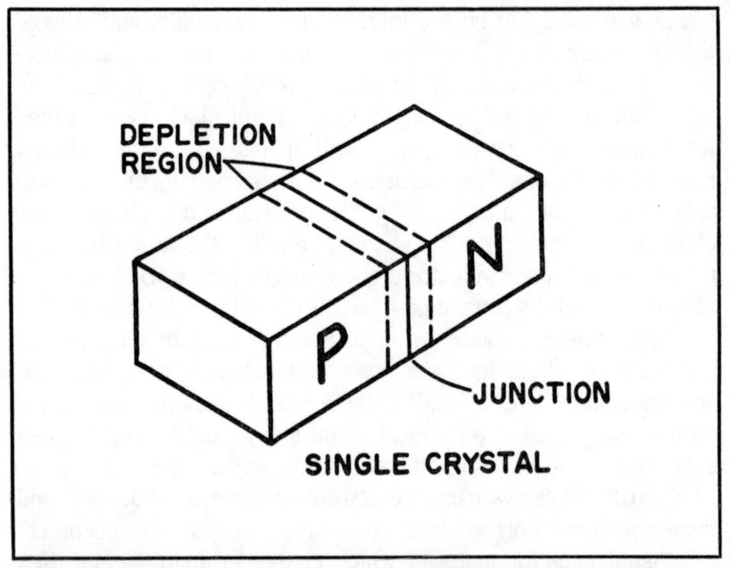

Fig. 4-8. Semiconductor diode consists of single crystal of either silicon or germanium, but two types of material. P-type has impurities which cause it to be short some electrons, and N-type has impurities which provide excess electrons. Excess electrons of N-type fit into vacant spaces in P-type, to create "depletion region" surrounding the "junction" where the two types meet. This depletion region corresponds to the space charge in a tube.

attracting more electrons out of the P-type material, creating new holes in a continuing process. These new holes will, in turn, be filled by electrons pushed through the depletion region from the N-type material. The flow of electrons through the depletion region is thus similar to that of electrons through the space between cathode and plate in a tube, and current flows in the external circuit.

If, on the other hand, we connect the battery so that the N-type material is made more positive (Fig. 4-10), then electrons from the battery's negative pole fill the holes in the P-type material, while the battery's positive pole attracts all the excess electrons from the N-type. Both the holes and the excess electrons are pulled away from the depletion region, and neither crosses through. With nothing going through the depletion region, very little current flows. This condition in a semiconductor is known as "reverse bias"; the opposite condition, when current flows freely, is called "forward bias."

In the vacuum tube, we inserted the grid between cathode and plate to exercise control over plate current. In a transistor, it's

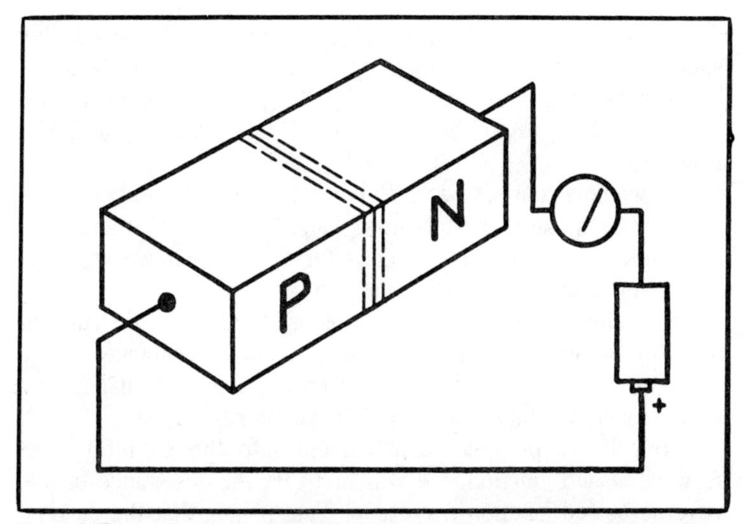

Fig. 4-9. Connecting battery with positive pole to P-type material and negative pole to N-type adds force from battery to make the depletion region become narrow. Electrons are attracted out of the P-type material by the battery, and in consequence more electrons are pulled through the depletion region. These must be made up for by fresh electrons from battery, and so current flows in external circuit. This is known as "forward bias" condition.

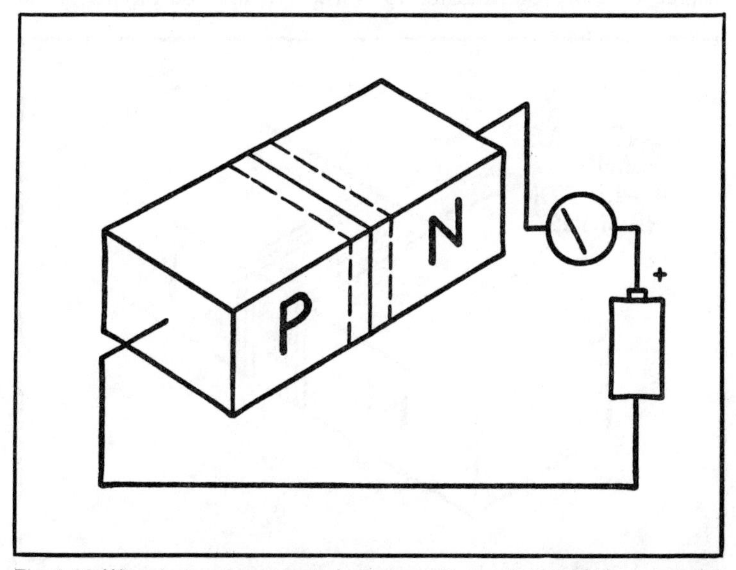

Fig. 4-10. When battery is connected with its positive pole to the N-type material and negative pole to the P-type depletion region is widened until the force across the depletion region equals the force applied by the battery. The two forces then balance each other so no current flows. This condition is known as "reverse bias" of the junction.

done differently. Two diode junctions are formed end-to-end in the same crystal, so that a a thin region of one type of material is between two regions of the other type (Fig. 4-11). If the ends are N-type, and the region between them P-type, we have an NPN transistor.

If we now connect two different batteries to these two diode junctions in the same crystal (Fig. 4-12), so that one of the junctions is reverse-biased and the other is forward-biased, we will find surprising interaction between the two.

The junction which is reverse-biased will let little current flow through and so will have comparatively high resistance.

The one which is forward-biased, on the other hand, will permit much current to flow and will have low resistance.

But if we put additional current into the circuit of the forward-biased junction, we will find that the resistance of the reverse-biased junction decreases! That means that the current flow through the forward-biased junction controls the resistance of the reversed-biased junction, and we have an amplifying device.

What happens is relatively simple if we dispense with most of the mathematics and double-talk which normally surrounds an explanation of how transistors work. We may be slightly less

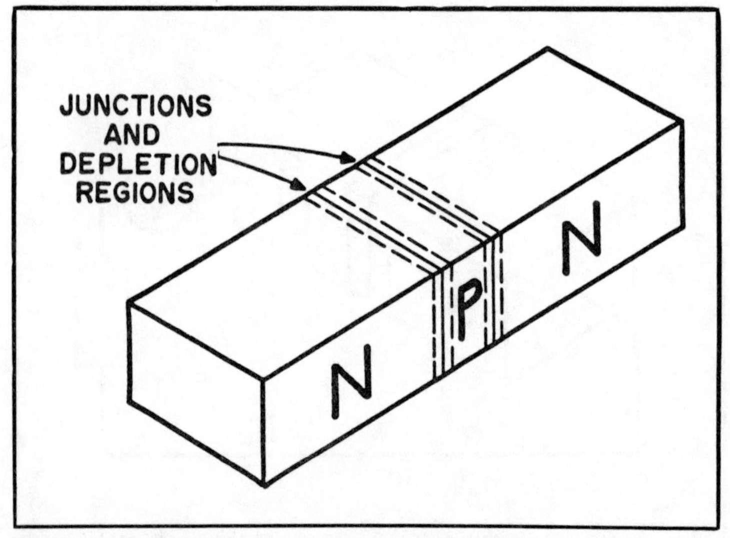

Fig. 4-11. Like a semiconductor, diode, a transistor consists of single crystal composed of two different types of material. But a transistor has two junctions (each with its own depletion region) rather than just one. Interaction between junctions, within the center area between the junctions, makes amplification possible within transistor.

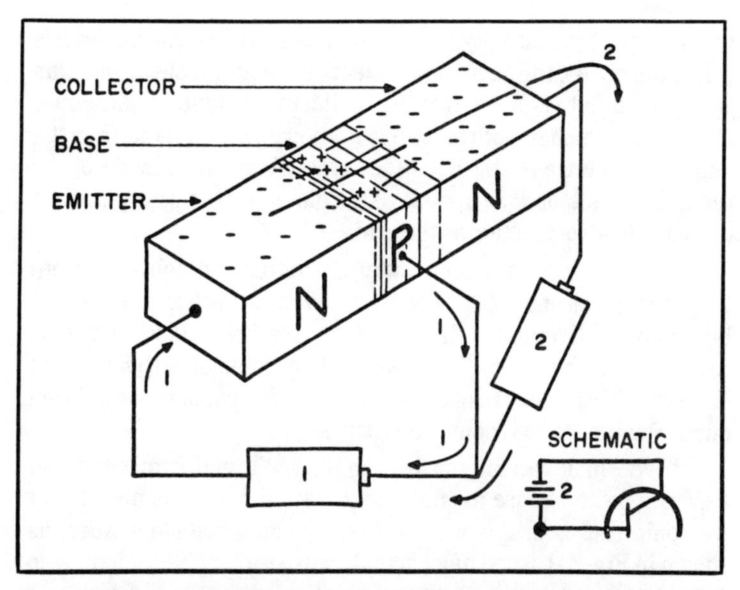

Fig. 4-12. Transistor operation is demonstrated here. Battery 1 provides forward bias to base—emitter junction; battery 2 puts base—collector junction in reverse-bias condition where it has high resistance. Most electrons forced through base—emitter junction by battery 1 go right through base into collector, reducing its resistance and permitting external current flow through outside loop. Note that this is simplified circuit, and would burn out any transistor so connected because nothing is present to limit collector current or base current.

accurate as a result, but at least we'll be understandable, and we won't be very far off the track.

When we put additional current into the forward-biased junction, this forces excess electrons out of the N-type material (which as Fig. 4-12 shows, is at one end of the crystal) with quite a bit of energy. Some of these have so much energy, in fact, that they shoot right on through the thin P-type region between the two junctions and reach the reverse-biased junction—from its P-type side, where the reverse bias has removed most of the holes into which they could fit. Thus they go right on through the reverse-biased junction as well as into the other N-type material, and there the positive voltage from the battery attracts them into the external circuit.

This means that more current flows in the reverse-biased circuit, which is another way of saying that its resistance is reduced.

The N-type material which is connected to the forward-biased junction is known as the **emitter** because it emits the electrons which do the transistor's job. The other region of N-type material is called the **collector** because it collects the emitted electrons. The thin region between emitter and collector, P-type in this case, which is common to both junctions is known as the **base**. Thus the emitter of the transistor does the same job as the cathode of the tube, the collector corresponds to the plate, and the base performs the same control function as the grid.

It would work the same way if emitter and collector were P-type material and the base N-type; the resulting transistor would be called a PNP unit and all the battery polarities would have to be reversed, but it would still amplify. Many common transistors, in fact, are PNP types; we used the NPN for this explanation to bring out the similarity to vacuum-tube actions.

If you look carefully at Fig. 4-12, you'll notice that the two separate batteries are in series with each other. This means that we could just as easily use one battery and a voltage divider, as shown in Fig. 4-13, and most actual transistor amplifier circuits do so.

In actual transistors, the emitter junction is usually built up differently from the collector junction, and that's why the emitter and collector leads are separately labeled on most units. Some, though, are built so that either outside region can act as emitter with the other as collector, and some will perform after a fashion if they are hooked up "upside down."

For the transistor to work properly, almost all the "charge carriers" (electrons in NPN units and the corresponding holes in PNPs) which leave the emitter must pass through the base into the collector. This action is determined by the transistor's construction just as the gain of the vacuum tube is determined by its geometry; modern transistors have about 99% of the charge carriers going to the collector and only 1% or less to the base.

You can see in the circuit of Fig. 4-13 that the emitter lead carries all the current of both the base lead and the collector lead. That is, the base current and the collector current added together equal the emitter current.

Two major characteristics are in common use to rate transistor gain. One, called "alpha," is the ratio of collector current to emitter current. Since at least some current must flow in the base circuit, the alpha of any transistor must be less than 1.0; the closer to 1.0 it actually is, the higher the transistor's gain. Almost all modern units have alphas greater than 0.90, and most exceed 0.99.

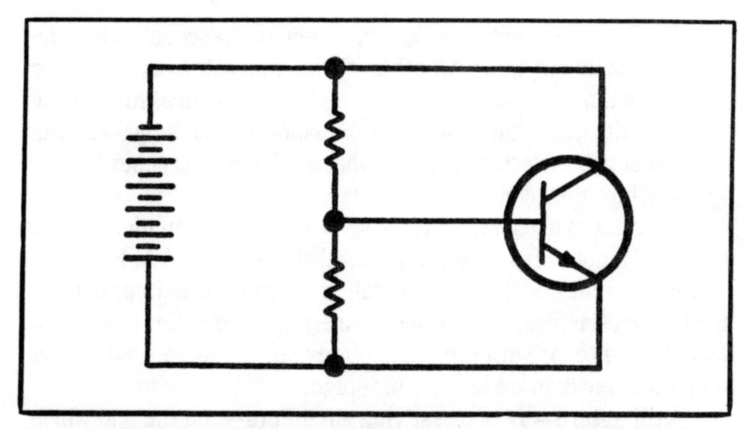

Fig. 4-13. Single battery plus voltage divider produces same result as two separate batteries. Lower resistor of voltage divider is often omitted, because base—emitter junction is forward-biased and effectively is short-circuit around lower resistor. Value of upper resistor is then determined by desired amount of base current, by Ohm's law.

Because the range of usable alpha figures is so small, the other characteristic is more widely used. It's called "beta," and is the ratio of collector current to base current. Thus if a 1 mA change in base current causes a 50 mA change in collector current, the transistor's beta is 50. Typical beta figures for present units range from 10 to 500, with 50—100 being the average range.

While "transconductance" could be and at times has been applied to transistors just as to tubes, it doesn't make much sense because the transconductance of most transistors is several thousand times that of typical tubes. Almost all transistor ratings are in terms of either alpha or beta.

Transistors are rapidly replacing tubes in many applications because they have several advantages. Since their action doesn't depend upon heat boiling electrons free, they don't require as much power as tubes. Since they work inside a crystal rather than in a vacuum, they don't need a special envelope. And since the action takes place at the atomic level, they are much smaller than tubes.

WHAT KINDS OF AMPLIFIERS ARE THERE?

Amplifiers can be classified in a great many different ways. One way, popular among those who write amplifier advertisements but not too meaningful for our purposes now, is according to the type of device which does the amplifying. That is, into "transistor" amplifiers and "vacuum-tube" amplifiers, with the "transistor" class also called at times "solid-state."

While it's essential to know, when you're considering the purchase of an amplifier, whether it uses transistors or tubes, we have seen that the two devices accomplish the same basic purpose in approximately the same basic manner—which makes this method of classification fall somewhat flat for our purposes here of learning how they work.

Engineers use another technique to classify amplifiers. They divide all amplifiers into three lettered classes, called A, B, and C, as well as some borderline areas called AB which is neither fish nor fowl. The classification is based entirely upon amplifier theory and has nothing to do with the kinds of devices involved—but it's far more detailed than we need at this stage.

Still a third way of classifying amplifiers—and the one which we shall use—steers a course between the non-theoretical and the completely theoretical, and divides all amplifiers into two groups, depending upon what they do to the signal in the process of amplifying it.

One group produces an output which is an exact replica of its input signal, only larger. It adds no distortion of its own. This group is known as "linear" amplifiers, often abbreviated simply to "linears."

Not all amplifiers produce outputs which are exact replicas of their inputs. Many distort the input signal in some manner; the distortion may be small, or it may be so extreme as to render the input signal virtually unrecognizable in most respects. Thus not all amplifiers are linear.

Those which are not linear form the other group, whether the departure from linear conditions is small or large. It's called the group of nonlinear amplifiers.

A top-quality hi-fi or stereo amplifier comes close to being a perfect example of a linear amplifier (although perfection in this, just as anything else, is impossible to achieve in practice). While any real amplifier is nonlinear to some extent, we call it "linear" if it's linear enough for our purposes, and a good audio rig produces so little distortion that expensive laboratory instruments are required to determine its departures from linearity.

The amplifiers used in most radio transmitters to boost the low power of the oscillator up to the higher level actually transmitted, on the other hand, distort the signal unmercifully in the process. They're designed to do so, because they can operate more efficiently that way and the distortion doesn't hurt anything. This is an example of a deliberately nonlinear amplifier.

The word "linear" in both these names comes from the technique often used to display an amplifier's performance. The input and output signals are used to draw a graph such as Fig. 4-14, which shows the output signal level for each value of input signal level for a pair of typical amplifiers.

The one on the left produces an output signal which is a perfect replica of the input signal, because no matter what level the input may have at any instant, the output is always just 16 times larger. Each variation of input level is thus reproduced, 16 times larger, at the output terminals.

That on the right, on the other hand, produces output 16 times as large as the input for very small input signals, only 8 times as large as the input for moderate input levels, and falls off to only 4 times for large input signals. Thus the output cannot be an exact replica of the input, because an 8-to-1 variation in input levels may be reduced to only a 2-to-1 variation of output levels.

Note that the line on the graph at the left is straight, while that on the right-hand graph is a curve, steep at the left and nearly level at the right. The word "linear" means simply that the input-to-output graph or "performance characteristic" of that amplifier is a straight line. A nonlinear amplifier's performance characteristic is not a straight line; it may be curved as in Fig. 4-14, or it may show one or more sudden sharp breaks.

One characteristic possessed by all amplifiers, linear or nonlinear, is gain. The gain of an amplifier is the measure of the amount of amplification it produces; it's figured by dividing output level by input level, and the resulting figure is the amplifier's gain.

That is, an amplifier which produces 10 watts of output power with 1 watt input has a power gain of 10. If 2 watts of input were required to produce 10 watts of output, the gain would be 10/2 or 5, while if 1 watt in produced 20 watts out, the gain would be 20.

The gain of any linear amplifier is constant. That is, the gain is the same regardless of input signal level. This is the only way that the output can be a replica of the input, because if the gain varies at all during operation, the output after the gain change is not the same as that before—although the signal at the input may not have changed.

The nonlinear amplifier's gain, on the other hand, changes depending upon any of several factors.

An important point to keep in mind is that any specific amplifier may be linear for small input signals and become nonlinear for larger ones. In fact, this is true of all actual amplifiers,

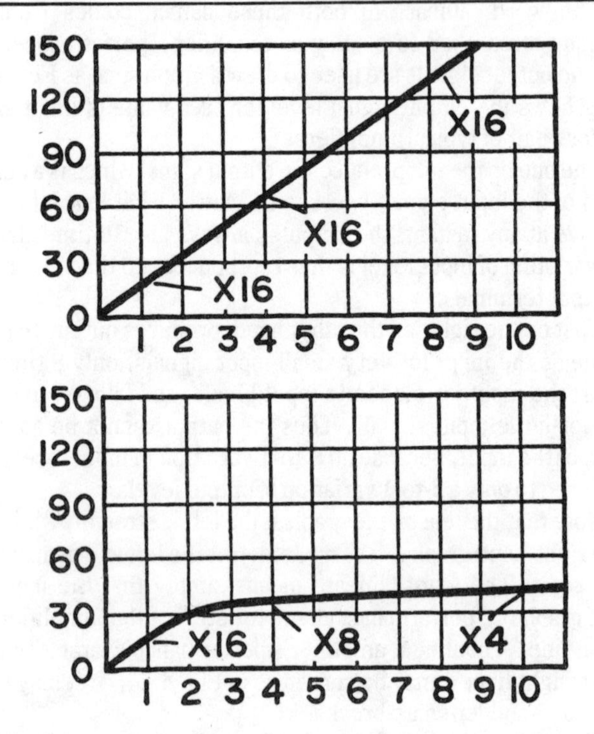

Fig. 4-14.

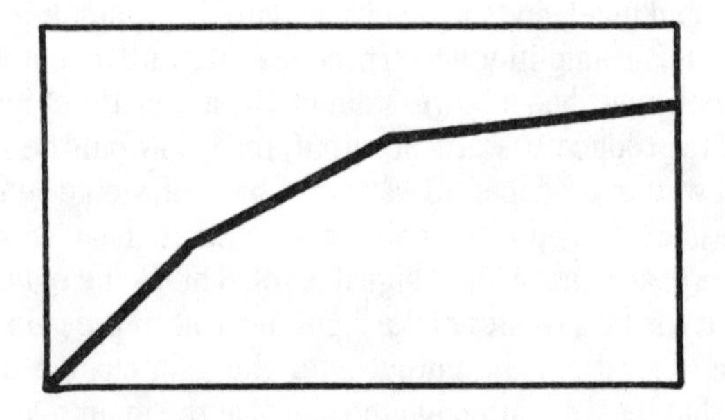

Fig. 4-15.

Figs. 4-14—4-18. These typical graphs of amplifier performance or "performance characteristics," show difference between linear and nonlinear operation. Only left-hand graph of Fig. 4-14 and part of Fig. 4-16 enclosed in dotted box are linear; rest are all nonlinear in one way or another although Fig. 4-17 is typical of most "linear" amplifiers in practice.

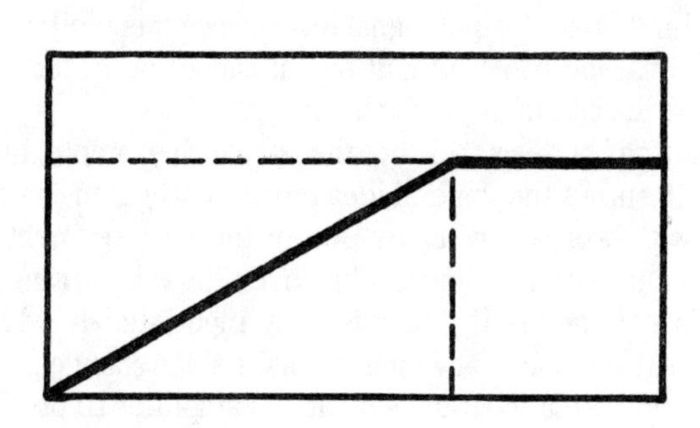

Fig. 4-16.

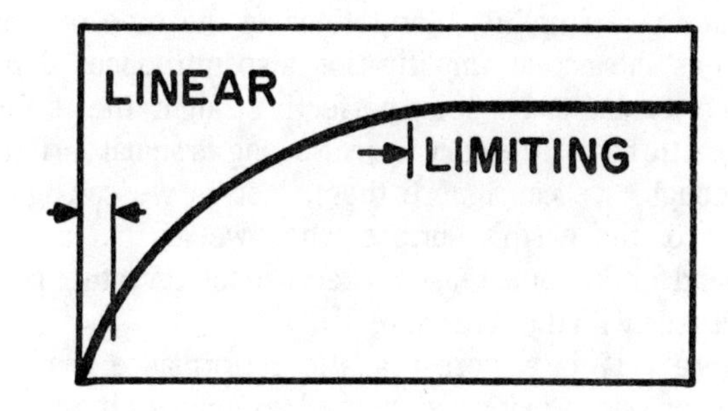

Fig. 4-17.

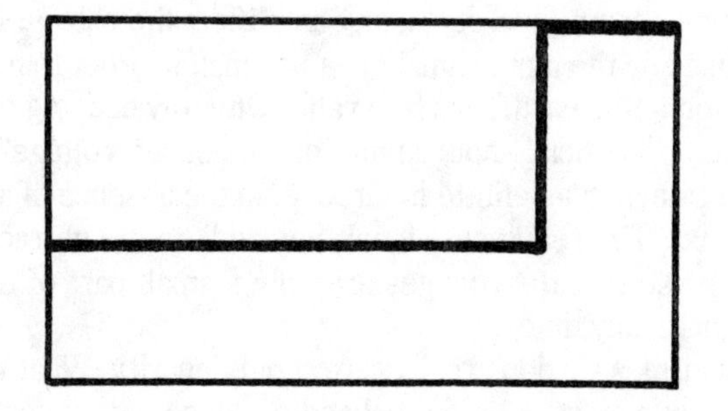

Fig. 4-18.

because the amount of power available from the power supply puts an upper limit on the output signal level. Once this limit is reached, any additional increase in input signal cannot be reflected in the output, and the amplifier cannot be linear.

"Limiting" of this sort is not the only such example, however. Figure 4-15 shows the general idea more clearly than can possibly be done with words. Over any one of the four segments of the curve, this amplifier is linear—but over the whole range of the graph, it's not, nor is it linear for any input signals which pass through the three points at which the line's slope changes.

It would take a most unusual actual amplifier to produce this performance characteristic, but limiting occurs in virtually all amplifiers. Figure 4-16 shows the effect of limiting upon an otherwise linear amplifier; the part of the characteristic enclosed in dotted lines is linear, but over the wider range, it's not.

In any real amplifier, limitations of the tube or transistor which does the actual amplification also introduces departures from linearity. Instead of being perfectly straight, the performance characteristic has a little curvature. So long as signal variations are small enough, we can ignore them, just as we can ignore the curvature of the earth's surface when we lay out a vegetable garden, and for the same reason—because the curvature is so small in comparison with the area we're using.

Figure 4-17 is a more realistic performance characteristic than any of our previous simplified examples. It includes the limiting caused by power-supply overload and the nonlinearity due to device characteristics, and the "linear" range suitable for use is indicated. Most actual "linear" amplifiers have performance characteristics similar to Fig. 4-17.

For maximum linearity in any amplifier, the change in input level caused by the input signal must be small in order to make the comparison to the earth's surface valid. One way of doing this is to introduce an artificial input signal or dc "offset voltage" at the input, to establish a definite input level in the absence of any true input signal. The real input signal then adds to or subtracts from this offset, so that the changes are only a small part of the total input signal at any time.

Sometimes we don't really care about linearity. What we want may be maximum gain, ability to handle high power, or some other feature. The rf power amplifiers in most transmitters are good examples of this kind of situation. We don't need to worry about distortion of the individual cycles of rf energy, because our tuned circuits will smooth them out again. We do, however, want to be sure that most of our power goes out the feedline to the antenna instead of merely warming up our shacks as heat dissipated by the tubes.

We can achieve the goal of high circuit efficiency by using the amplifier as a "switch." This is an example of a very special kind of amplifier which produces no output at all until its input signal is above a "threshold," then produces maximum or near-maximum output until input falls back below threshold again. Figure 4-18 is a typical performance characteristic for such an amplifier; you can see that it's anything but linear.

This type of amplifier wastes very little power, because when the input signal is below threshold and no output current is flowing, no power is being taken from the supply. When the situation is reversed and near-maximum current is flowing, the tube or transistor must have as low a resistance value as it can achieve—and this means that the least possible amount of power is being dissipated as heat in the resistance of the tube or transistor. Most of the power taken from the supply shows up in the output signal.

Such an amplifier is the most efficient of all amplifiers so far as power dissipation is concerned. That is, it loses less of its dc input power (from the power supply) as heat than does any other kind. Since the power dissipation is one of the major factors limiting the application of any specific tube or transistor, this means that any specific amplifying device can produce more output power in this type of circuit than it could in any other type of amplifier.

The power efficiency of this kind of circuit works for us in two ways. It's economical of power, which is always a bit difficult to come by, and it's economical of tubes or transistors, because we can use a smaller and presumably less costly device to get the same output power level (as compared to a linear). That's one of the reasons why this type of amplifier is used for rf in transmitters, whenever we can get by with it.

For some types of signals, such as SSB or TV video, we can't accept the distortion of the superefficient nonlinear circuits, and must make do with the relatively inefficient linear arrangement. And in audio, where we cannot tolerate distortion of individual cycles, we must always be concerned with linearity.

HOW DO AMPLIFIERS AMPLIFY?

Now that we've looked at the two major groups of amplifiers, linear and nonlinear, and discovered where and why each kind is used, let's see how any amplifier does its job of making an input signal stronger.

We saw earlier that the only way an amplifier can add power to its input signal is to take power from some external supply, and control the amount of power in its output circuit according to the input signal level at each instant.

One of the simplest ways of controlling the amount of power in any circuit is to vary the resistance present in that circuit, because a circuit with low resistance will permit more power to be used than will one with high resistance.

And that's just how transistor and vacuum-tube amplifiers operate.

Either a transistor or a tube can be thought of as a resistor which has the unusual property of changing its resistance by purely electrical means, as contrasted to the mechanical change of resistance we achieve by turning the shaft of a rheostat or a potentiometer, or by moving the slider on an adjustable resistor.

In the vacuum tube, the resistance which is changed is that between the plate and the cathode. In the transistor, the collector-to-emitter resistance changes. This permits current flow in the plate or the collector circuit to be the output signal.

The resistance is changed by a signal applied between grid and cathode in the vacuum tube, or between base and emitter in the transistor. Thus the input signal normally goes to the grid or the base.

Charts are available for just about all tubes and transistors used as amplifiers, which show plate or collector current versus plate or collector voltage for specified grid or base voltages. Though these charts don't show resistance as such, the current and voltage values shown on them determine resistance, and so you can think of them as indicating the resistance changes also.

Figure 4-19 is a typical plate-characteristic chart for a triode tube. It tells us many important things about not only the particular tube type to which it applies, but all tubes. For instance, higher current flows when the grid is positive than when it is negative—but the saturation point is reached with very low plate voltages under such a condition, and the grid then loses control. A saturated tube cannot amplify.

As grid voltage becomes more negative, plate current decreases. The cutoff condition is easy to see on this chart; plate current is nearly zero over a wide range of plate voltage.

Earlier, we saw that one way to increase linearity was to keep the change in input level caused by the input signal to only a small

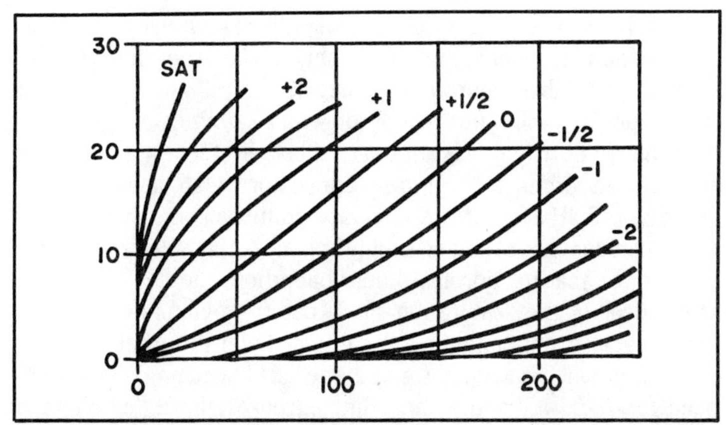

Fig. 4-19. Typical set of characteristic curves for triode tube shows how plate current, plate voltage, and grid voltage relate to each other. Note that raising grid voltage from 0 to 1 at 100V raises plate current from 10 to 21 mA; to produce this much change in plate current by charging plate voltage alone would require an increase to about 170V, or 70 times as much. Thus this tube has amplification factor of about 70.

part of the total input level, by introducing a dc offset voltage into the input signal. This voltage is called bias.

For instance, if our input signal has a 1-volt peak-to-peak swing, the signal level itself would swing from +0.5 volt to −0.5 volt. If we simply applied this signal to the tube whose characteristics are shown in Fig. 4-19 (at a constant plate voltage of 100), we would have a plate current of 16 mA at the positive peak of the input signal, 10.5 mA when the input signal passed through zero, and 7 mA at the negative peaks. The change from zero input to positive peak would be 5.5 mA, while that from zero to negative peak would be 3.5 mA. Since these are not equal changes in output, although the input changes were equal, the amplifier would not be linear.

However, if we apply a dc bias of −0.5 volt to the tube's grid, the input signal will then cause the grid voltage to vary from 0 at positive peaks (+0.5-volt signal and the −0.5 volt bias canceling each other) down to −1 at negative peaks. Plate current then is 10.5 mA at positive peaks, 7 mA with zero input, and 3.5 mA at negative peaks. The change from zero to either peak is 3.5 mA. Since the output changes equally for equal changes in input, the amplifier is linear for this signal and bias.

The circuit designer must juggle the expected signal levels, plate voltages, and bias values for each amplifying device in his circuit in order to produce the proper compromises between gain, linearity, and power consumption.

While we used a vacuum-tube amplifier as our example, transistors operate in much the same fashion. The major difference is that vacuum tubes operate with negative bias, normally, while transistors are at the cutoff condition with zero or negative bias and so require positive bias (forward bias on the base—emitter junction) in order to operate. Transistors also come in two polarities, PNP and NPN, while vacuum tubes come in only one, which corresponds in power-supply hookup to the NPN transistor.

Notice that for normeal linear operation, the designer must steer clear of the "saturation" and "cutoff" regions because the input signal loses control of the output level in either of these limiting areas. However, for the special "switching" amplifiers used for rf power and some other purposes, both "cutoff" and "saturation" are the useful areas, and the in-between "linear" region is avoided. In this case, the device is normally biased to cutoff, and the input signal is made large enough to not only overcome the bias on its positive peaks, but drive the device over into the saturation region thus achieving lowest device resistance and wasting the smallest possible amount of power.

The variable-resistor viewpoint of amplifier operation we've presented here is one you won't find in many theory texts; the texts are, for the most part, written by and for engineers who tend to stay with traditional approaches to such things. It does, however, bring out the similarities between tubes and transistors in a way unmatched by the traditional approach, and can serve every other purpose as well (in fact, it even makes the traditional approach easier to comprehend).

Chapter 5
Rules and Regulations: A Discussion

"Law 'n' order" seems to be a loaded phrase these days, which (depending upon one's personal biases) may be used to excuse a multitude of injustices or to rationalize the actions of a vigilante gang. These politico-sociological questions are far outside the scope of our study course for the amateur radio license, but the subject of "law 'n' order" itself is not. Regardless of one's political or social leanings, it should be obvious that *some* form of control and authority is necessary when it comes to radio transmissions. After all, there's only a limited amount of room in the rf spectrum. Without some controlling force, the interference which would result if anyone who wanted to were able to put a transmitter on the air would make radio unusable as a medium of communication, entertainment, or any other public benefit.

The lawmakers of the world recognized this many years ago, while the art of radio was still in its earliest infancy, and almost without exception they took action to gain (and keep) control of the privilege of operating a radio transmitter. We in the U.S. were relatively lucky; in many nations, radio communication was taken over as a government monopoly (Japan, England, Germany, to name only a few) and the activities of private radio operators, both amateur and business, were restricted much more severely than here.

But here, the federal government declared its control of the airwaves, and in several court cases, established that individual states had no control over radio facilities. During World War I, the Navy Department controlled radio. Afterward, the Commerce Department took it over and administered it until the Communication Act of 1934 established the Federal Communications Commission as a separate part of the government's executive branch, with full control over civilian radio operations.

The FCC's authority stems from that Communications Act, which has been amended several times in the ensuing years. The Commission consists of seven commissioners, each appointed by the President and approved by the senate, together with a moderately large staff. The commissioners make the policies, and the staff carries them out (although in practice, many rules changes are initiated by staff action; however, no change can become effective until so ordered by the commissioners).

The complete text of the Communications Act in its present form would occupy many pages; even those parts of it which directly affect the amateur radio service would take more space than we have available, to quote in full. A number of points, however, are so important that they should be noted:

Section 301, for instance, provides that the Commission may grant the "use" of channels for radio communication under "licenses," but may not confer "ownership" of them, and requires that the licenses be issued for "limited periods of time"; it then goes on to require that "any apparatus for the transmission of energy or communications or signals by radio" be operated only "under and in accordance with this ACT and with a license in that behalf granted under the provisions of this Act."

Since the Communications Act is a part of the body of both civil and criminal law of the federal government, this means in plain language that it's against the law to operate an unlicensed transmitter. You may have heard that the FCC has no jurisdiction over anyone but license holders...and that's quite true. The FCC does, however, operate the monitoring stations necessary to enforce the licensing provisions of the Act; when they locate an unlicensed station, it may be an FCC man who first calls politely to warn the operator...but if it's necessary to get tough the Commission just turns everything over to the Department of Justice, whose enforcement branch (the FBI) handles the actual arrests and charges.

Section 501 establishes criminal penalties for violation of the Act. Stripped of the legal terminology, the penalty for violation of the Act itself is $10,000 fine, one year's imprisonment, or both (these are maximums; no minimum is set). Violation of FCC rules and regulations (as opposed to the Act itself) may be punished by a forfeiture of $500 for each day of violation, or by license revocation, or both.

Section 505 contains the famous "secrecy" clause, which says in essence that you cannot legally repeat to anyone (or make a profit for yourself) anything you hear on radio *except* those items broadcast by broadcasting stations, or relating to ships in distress.

The Communications Act itself is rather broad in scope, dealing as it does with delegation of authority from Congress (through its constitutional power to control and regulate interstate commerce) through the executive branch to the FCC itself. Most of the day-to-day operations of all radio stations, commercial and amateur alike, are controlled by the Rules and Regulations of the FCC (authorized under Section 303 of the Act).

These Rules and Regulations are, themselves, quite bulky. They are published by the Government Printing Office in a number of volumes because of their bulk, and are divided into "parts"—each "part" dealing with a separate class of radio service. The amateur radio service is regulated by Part 97, which is reproduced in Appendix C of this book.

Part 97 is divided into several lettered subparts; only subparts A through E are of major interest to the would-be holder of a Novice license. Subpart A covers "general" subjects, primarily setting forth the phrases used elsewhere in Part 97; subpart B covers amateur operator and station licenses; subpart C regulates technical standards; subpart D prescribes operating requirements and procedures. Finally, subpart E sets forth prohibited practices and administrative sanctions.

The Novice license is established by a portion of subpart B, and the privileges of the Novice operator are spelled out in paragraph 97.7.d as follows:

(1) The power input to the transmitter final amplifying stage supplying radio frequency energy to the antenna shall not exceed 250 watts, exclusive of power for heating the cathode of a vacuum tube(s).

(2) Radio telegraphy is authorized in the frequency bands 3700—3750 kHz, 7100—7150 kHz (7050—7075 kHz when the terrestrial location of the station is not within Region 2), 21,100—21,200 kHz, and 28,100—28,200 kHz, using only Type A-1 emission.

The term of the license is five years, and it is renewable. Formerly, no one who had ever held an amateur license was eligible for a Novice license, but this has been changed so that a former ham can apply for another Novice license if he has not held any FCC license in the previous 12 months.

The regulations discuss and describe "operator" and "station" licenses separately, and they are treated officially as separate licenses. To be eligible for a station license, though, anyone must hold an operator license as well (except for a few special instances involving club stations on government property), and in most cases any one operator is limited to holding one station license in any single call area. As a result, you get only *one* piece of paper which is *both* licenses; being issued together, they expire together.

Should you fail your first attempt at any amateur exam, you must wait 30 days before trying again. There is, however, no limit to the number of times you may apply.

Besides the establishment of the license class and the description of its privileges and limitations, subpart B contains at least one other item of interest to would-be Novices since it may be part of the examination; this item is the limitations on the antenna structure (paragraph 97.45).

The reason for limiting amateur antennas is to prevent them from becoming hazards to aircraft, unlike the 20 ft. limit on CB antennas which is meant to restrict their operating range.

As a result, the limitations on amateur antennas are much less strict. No *absolute* limit is established, but prior FCC approval is required for installation of a ham antenna more than 170 ft. tall, or when the proposed antenna is more than 20 ft. above ground and is within 4000 ft. of an airport runway. In the latter case, one foot of antenna height is permitted for every 200 ft. of distance from the runway before approval is required.

In practice, few ham towers exceed 100 ft. in height, and the requirement for prior approval consequently does not apply to most stations.

The major points of interest for Novice applicants in subpart C are the requirements for "pure and stable" signals and for frequency measurement.

Paragraph 97.73 establishes the requirements for signal purity and stability. "Spurious radiation" must be reduced or eliminated "in accordance with good engineering practice." This means that every effort must be made to prevent harmonics from being transmitted; key clicks, parasitic oscillations, and "backwave" (key-up signals) are also spurious radiations, but harmonics provide the major part of the problem.

Frequency must be as constant as the state of the art permits; this usually offers little problem, but "chirp" when keying may still cause violations of this regulation. Finally, the plate power supply must be "pure" to avoid transmitting "hum" modulation.

Paragraph 97.75 requires that the licensee of any amateur station must provide for measurement of the emitted carrier frequency of his transmitter, and must establish a procedure for making such measurements regularly. It goes on to require that the apparatus used to make this measurement must be independent of the transmitter itself, and of sufficient accuracy to insure that operation is restricted to the band or bands authorized for that transmitter.

This sounds impressive, but in practice a well calibrated receiver is usually sufficient, together with a "frequency marker" which is simply a crystal oscillator used to generate a standard frequency (normally 100 kHz). The "regular procedure" requirement is satisfied by checking to be certain you are in the band every time you change the transmitter frequency.

Subpart D covers "operating requirements and procedures," and knowledge of its provisions is important. Typical questions you may encounter from this portion of the rules include "Who may operate an amateur transmitter?" and "When can you broadcast?"

The rules make a clear distinction between "operating" a station and "using" a station. "Operation" includes actual control of the transmitter, while "use" includes such things as permitting someone to speak over a phone station, or operate the keyboard of a RTTY installation. Only licensed amateurs may *operate* a station, but anyone may "use" it so long as control is retained by a license amateur.

Any licensed amateur may operate any amateur station subject to several restrictions. The operator must remain within the rules which pertain to his *own* class of license, and *also* those for the class of license held by the station's owner. Thus, a Novice operator may operate a station licensed to an Extra class operator only in the Novice bands, observing the 250W limit. The Extra class, privileged to use 1000 watts of power at his own station, is restricted to the Novice rules when he operates the Novice's station.

The distinction between "operator" and "station" licenses becomes important in subpart D. The *original* operator license must be in the operator's personal possession whenever he is operating, whether at home or away; the station license, on the other hand, must be posted in "a conspicuous place" while the station is being operated. Since both licenses normally are on the same piece of paper, it's fortunate that paragraph 97.85 permits a photocopy of the station license to be posted, thus allowing the operator to keep the original on his person while meeting the posting requirement.

When operating an amateur station, the station must be identified by transmission of its assigned call sign at the beginning and end of each single transmission or exchange of transmissions, as well as at intervals not greater than every 10 minutes during such an exchange. In addition, at the end, you must identify the other station.

If you are operating "portable" or "mobile" (about which more later), you must provide additional identification. For the Novice, this amounts to the fraction-bar CW character (dadididadit) followed by the call-sign area number in which you are operating. For instance, K4IPV would identify when either portable or mobile in call area 2 as K4IPV/2. In call area 5, it would be K4IPV/5, etc.

In general, a ham station may be used only to communicate with other amateur stations. In emergencies, it may communicate with *any* licensed station. Noncommunications uses such as radio control, experiments with antennas, and the like are also permitted, but "broadcasting" is prohibited.

Broadcasting is interpreted as the transmission of one-way communications, and no amateur station may do so. Four special cases of one-way communications which are not considered to be broadcasting are spelled out in paragraph 97.91, however; these are emergency communications (including practice drills), information bulletins consisting solely of subject matter having direct interest to the amateur radio service as such (primarily permitting the W1AW information broadcasts but open to any ham), roundtable or net operations where each station is transmitting to several others at the same time, each taking a turn at transmitting to the rest, and code practice transmissions (which must be clear language, since no codes or ciphers may be transmitted by ham stations).

Any amateur station may be operated "portable" or "mobile." "Mobile" installations are those in any vehicle which may be and normally are operated while the vehicle is in motion. "Portable" installations are any others at locations other than the permanent site specified on the station license; a "portable" station need not be capable of being moved conveniently. For instance, an operator having one station at home and another at work, would normally consider the one at work as "portable." Sometimes, an operator becomes attached to his callsign and must move to another call area; if he retains a legal residence in the old call area and keeps his permanent station license at that legal residence, it's perfectly all right for him to operate as a "portable" in the new call area with his old call sign so long as he meets the requirements for giving notice of portable or mobile operation.

These notice requirements apply only if the operator expects to be absent from his permanent transmitter location for more than 15 days; thus, a mobile installation used all over a call area needs no notice so long as the operator returns to his fixed location at least once each 15 days.

If the absence exceeds 15 days, or is likely to do so, then advance written notice must be given to the engineer in charge of the radio district in which operation is intended. The notice must include the licensee's name, call sign, fixed transmitter location, portable location or mobile itinerary as specifically as possible, the dates of beginning and ending of such portable or mobile operation, and the address at which the licensee can be reached during that time. For mobile operation, full identification of the vehicle in which the station is installed is also required. Once such notice is given, it must be renewed once each year if the operation continues over a year, or whenever any change occurs in the required information.

One more point spelled out in subpart D is the requirement for a written log of station operation.

We've already seen one prohibition from subpart E—the rule that says no amateur station may engage in any form of broadcasting (paragraph 97.113). Several others are also important.

Paragraph 97.119 prohibits the transmission of communications containing "obscene, indecent, or profane words, language, or meaning." Recently, at least one FCC commissioner has questioned the enforceability of this rule in view of the court rulings concerning obscenity, but the rule *is* still on the books and *may* be enforced.

Paragraph 97.115 prohibits the transmission of music by any amateur station. Single audio tones may be transmitted for test use.

Also prohibited are willful damage to radio apparatus (97.127), false or deceptive signals and use of fraudulent call sign (97.121), deliberate interference with other stations (97.125), unidentified communications or signals (97.123), helping others obtain licenses by cheating or fraud (97.129), and accepting any form of compensation, direct or indirect, paid or promised, for transmitting or receiving messages (97.112).

In addition to the Communications Act and the FCC Rules and Regulations, amateur radio is subject to provisions of several international agreements. The most important of these in day-to-day operation is Article 41 of the Geneva Radio Regulations, which governs international amateur radio contacts and provides that no such contacts shall be made on behalf of anyone other than the amateurs involved. Such communications are limited to messages of a technical nature and to personal remarks not important enough to justify commercial communications, and may be prohibited altogether if one of the countries involved objects. Several countries in the world have filed such objections, and their hams cannot talk to hams of other nations. Some of the provisions of Article 41 can be modified by special agreement between the nations involved, and the U.S. has special agreements permitting third-party message traffic between U.S. hams and hams in some other countries. These agreements are modified from time to time, and the count changes; at this writing, 22 countries permit such communications with the U.S.

This has been a general look over the regulations which a would-be Novice licensee must know in order to pass the exam, but we cannot emphasize too strongly that rules and regulations *do* change from time to time and therefore this information may not be completely valid by the time it reaches you. (Basic theory, on the other hand, changes rarely if at all and can be depended upon to remain true over a period of many years.) Because of this, it's always a good idea to check over the very latest FCC Rules and Regulations, Part 97, before taking any license exam in the amateur service.

Chapter 6
Transmitters

Gaining the ability to (legally) operate a radio transmitter is the whole point of obtaining a ham license, since any other radio equipment imaginable may be used without any type of license. It's not surprising, therefore, that nearly a quarter of the questions on the Novice class study list furnished by the FCC deal in one way or another with transmitter operation.

Now that we've become familiar with the basics of electronics theory, the operation of amplifiers, and the rules and regulations which govern the operation of Novice stations, we're ready to turn our attention to this all-important subject of transmitters.

Transmitter operation can be divided into categories for study in several ways: we could divide it into "theory" and "practice," by the components which constitute a typical transmitter, or in accordance with the "links" in the transmitter-to-receiver chain. We'll do a little of each, and group our points for discussion into "Basic Transmitter Theory," "How Signals Propagate," "Typical Transmitter Components," and "Practical Problems."

Some of these areas may appear to overlap—and they do, just like everything else in electronics. The deeper you delve into this fascinating subject, the more you will become convinced that no part of it stands isolated from any other part. Everything fits together so closely and in such an interlocking fashion that any attempt to isolate a part for discussion in a study course such as this must be rather artificial. We'll take this into account, and cross back and forth between subjects as we go along.

BASIC TRANSMITTER THEORY

The purpose of a radio transmitter is to make communication possible by radiating energy. More precisely, the transmitter develops the energy in a form which can be radiated, and the antenna system does the radiating.

Note that the transmitter's purpose is *not* "to communicate." Communication involves the transmission of some type of signal which has meaning assigned to it, and only the operators of a communications system can assign meanings. The equipment itself merely makes communication possible.

To accomplish this purpose, the transmitter must produce energy at its output in a form which can be radiated, and must also provide some means by which the energy can be "modulated" to produce a signal.

The simplest form of this "modulation" is to turn the transmitter on and off in a pattern which has prearranged meaning. When a transmitter is operated in this fashion, we don't usually say that it's being modulated. Instead, we call it a "CW transmitter" and the on-off pattern is usually called "code." However, this is just as much a modulating technique as is the transmission of video by a TV station, or voice by an Extra class operator. It's just a different type of modulation.

So a transmitter must have something which produces energy, and something which modulates it. In addition, most transmitters have some type of power supply to provide the raw energy input. The rig itself cannot create the energy which it sends out; rather it converts energy from some basic form into electromagnetic energy.

Now we're into the subject of "typical transmitter components," so we'd better do some backtracking and find out just how the energy is produced.

The energy must be in a form which can be radiated, which means that it must be alternating current at radio frequency. The rules and regulations specify the frequency bands within which Novice stations may operate, and the final output of a Novice transmitter must fall within the legal limits of the Novice bands.

So, starting with dc power, how do we get this rf output? Normally, we use an oscillator to generate rf from the dc, and then use one or more amplifier stages to bring the power level of this rf up to the desired value. FCC regulations limit us to 250 watts input at the final amplifier stage.

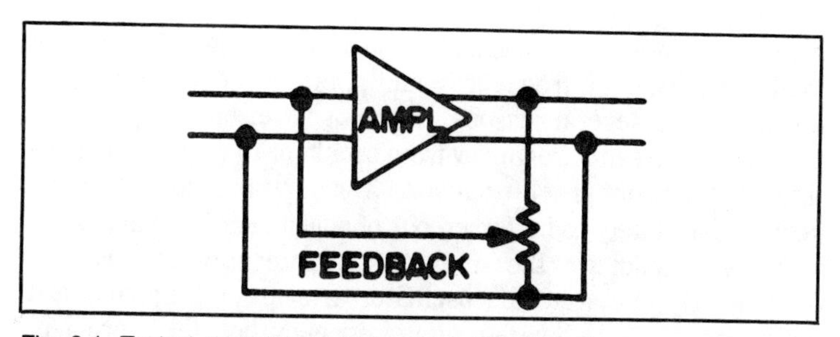

Fig. 6-1. Typical oscillator consists of amplifier with a portion of its output connected back in "feedback" arrangement to provide its own input. When phasing and amount of feedback are correct, no additional input is necessary. Rf oscillators also include some means for controlling frequency, not shown here.

A typical oscillator (Fig. 6-1) is simply an amplifier in which a part of the output signal is fed back to the input. When this is done, any fluctuation in either input or output—such as is caused by the supply voltage changing when power is first turned on—is amplified. A part of the amplified change comes back to the input, where it is amplified again. The process continues like a dog chasing its tail; no other input is necessary.

In addition to this simple feedback, however, the circuit must contain some sort of "resonator" to control the frequency at which oscillation occurs. Otherwise the signal produced will most likely not be usable for communication purposes. This resonator may be either a tuned circuit or a quartz crystal. The quartz crystal acts like an especially excellent tuned circuit, permitting energy to pass through at only one frequency and blocking all the rest.

Because of this selective action by the crystal or the tuned circuit, the only energy which can make it through the feedback loop is that which happens to be at the frequency of the resonator. If everything is properly adjusted, though, this is enough to produce an output, and so the oscillator merrily continues to chase its tail, at a frequency determined by the resonator.

That, in a nutshell, is how we convert dc power to rf energy. The output of an oscillator, though, is never very great. If we attempt to drive the circuit with more power, the result is that the crystal or resonating tuned circuit heats up and fails to operate properly. Crystals may even be shattered by excessive drive. Because of this, we normally use several stages of amplifiers to boost the power level.

While we're boosting the power, we can also change the frequency. An amplifier used to change the frequency to some higher multiple of the original frequency is called a frequency multiplier. Use of frequency multipliers is common in ham equipment for several reasons. If a transmitter operates straight through on the same frequency from oscillator to output, it's more likely to break out in self-oscillation from unwanted feedback, and also is more likely to be forced out of adjustment by unavoidable variations in loading as the wind blows the antenna or rain coats the feedline. By operating the oscillator at a lower frequency and multiplying up to the desired output frequency, both these problem areas can be minimized.

In addition, most of the ham bands are harmonically related to each other, so that by using frequency multiplication it's possible to make one crystal serve on several different bands.

The only difference between a frequency multiplier and a normal rf power amplifier is that the input and output circuits of the multiplier are tuned to different frequencies (the output being tuned to some exact multiple, or harmonic, of the input frequency), and the operating conditions of the amplifier are adjusted to increase the harmonic content of its output signal. This means that the proper operating conditions for a frequency multiplier are almost exactly opposite (in the harmonic-generation sense) to those for a normal rf power amplifier, because in the multiplier we want to produce harmonics, and in the normal amplifier harmonics are exactly what we do *not* want in the output.

When we get the frequency we want, we must use at least one normal amplifier stage to help prevent unwanted harmonics. This would indicate at least three stages in even the simplest transmitter making use of frequency multiplication, except that we can often perform the multiplication in the oscillator stage itself.

Now that we have the power we want at the frequency we want, we can think about modulating this energy to carry a message. Since Novice operations are restricted to the simplest form of modulation, CW, all we need do is turn the power on and off by means of a key. As we shall see in our discussion of practical problems, this is not necessarily as simple as it sounds. The frequency of an oscillator varies slightly when its operating voltages change, and the difference between "on" and "off" is quite a large change. This means that if we key the oscillator, the signal frequency will change on each operation of the key, giving rise to what is called "chirp." In many cases the easy way around this problem is to let the oscillator run at all times except when listening, and key only the amplifier stages.

Even though, as a Novice, you will be restricted to CW operation, its necessary to know at least a little about some of the other forms of modulation, if for no reason other than to be able to avoid getting them into your signals by accident.

Amplitude modulation, the form normally used for radio broadcasting and in ham phone stations, is accomplished by varying the signal power while keeping the frequency unchanged. Frequency modulation or FM, on the other hand, is done by changing the signal's frequency and keeping the power unchanged. Figure 6-2 shows the key differences.

While rather complex circuits are necessary in order to put voice on a radio signal by either AM or FM, it's easy to get AM or

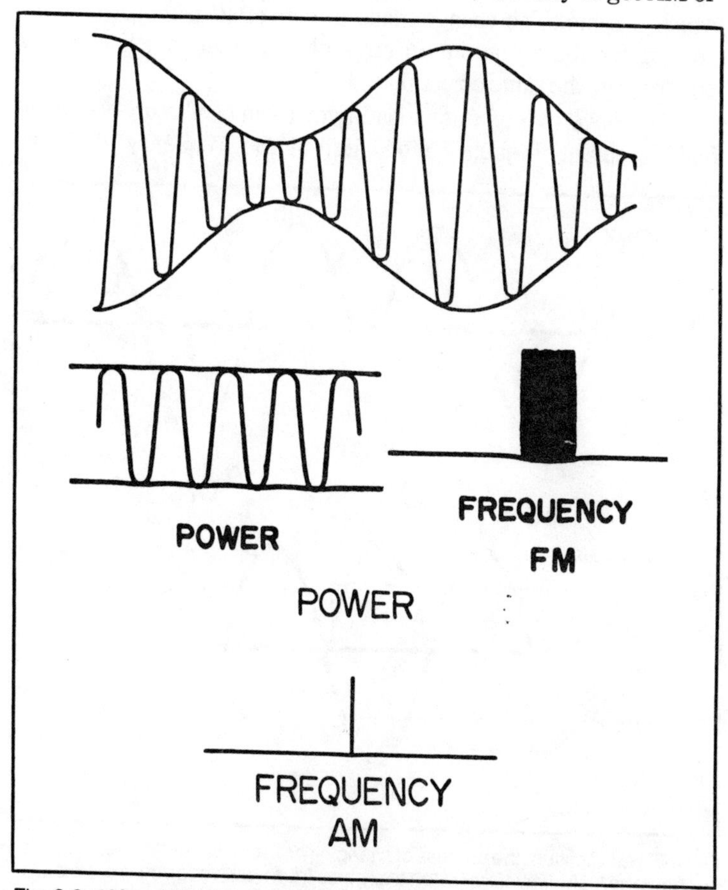

Fig. 6-2. AM and FM differ as shown here. With AM, information is carried by change in power level of signal and frequency remains constant. With FM, power remains constant and information is carried by changes in frequency.

FM of nonvoice signals by accident. Anything which causes the frequency of the output signal to vary is FM. We've already met one example—the chirp caused by keying the oscillator. Similarly, anything which causes the output power to vary causes AM.

amplitude

Power Supply Filtering

One of the biggest causes of unwanted AM and FM signals which are supposed to be CW is inadequate filtering of the power supplies. To get the power to operate the transmitter, we usually transform ac from the normal power outlet to the appropriate voltages, then convert it to dc. At this point, the dc is full of "ripple" (Fig. 6-3) and varies in strength just as did the ac from which it came. The change made to convert it from ac to dc did nothing for the variations in strength, but merely made all the electrons go the same direction.

Were we to feed this pulsating dc to an oscillator, we would find its output frequency modulated at the frequency of the ac

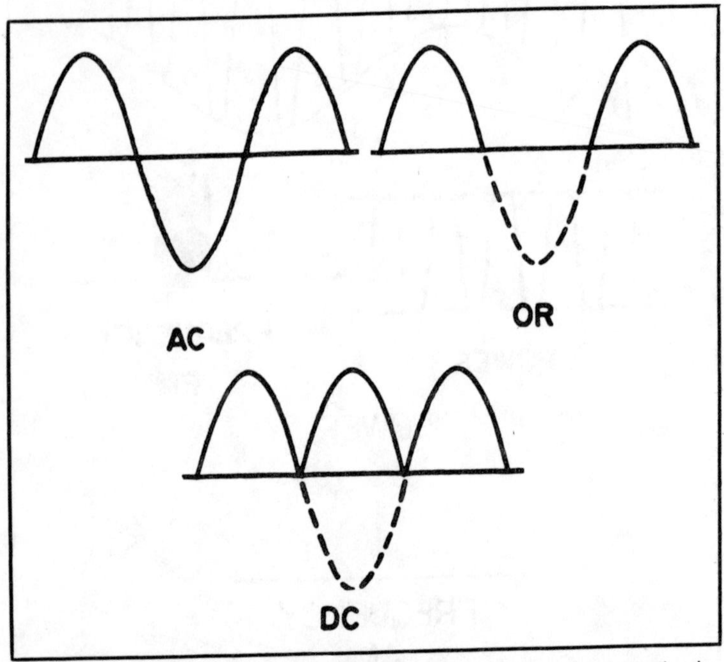

Fig. 6-3. Rectification (the process of changing ac to dc) leaves "ripple" in the dc as shown here. Ac waveform at left alternates from positive to negative polarity. Dc waveforms at right are always positive but vary from peak voltage to zero volts. Frequency of ripple component is same as that of ac when half-wave rectifier is used (center) or twice that of ac with full-wave rectification (right).

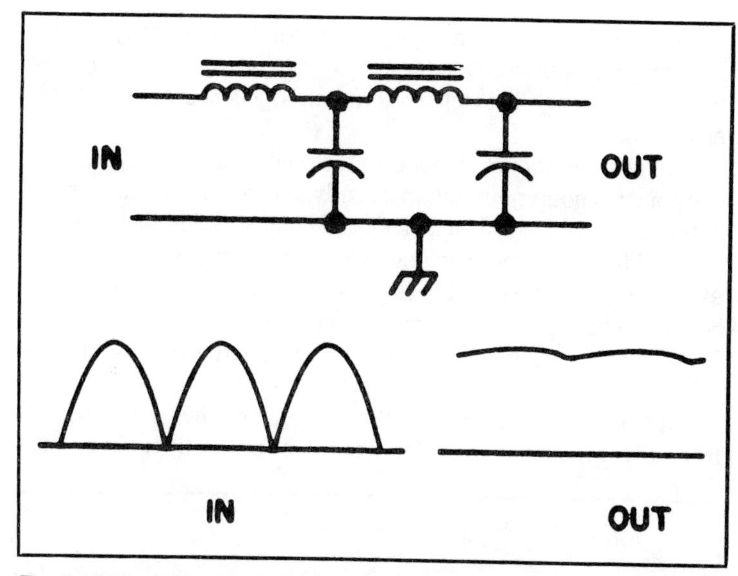

Fig. 6-4. Filter in power supply removes ripple. Inductors oppose changes in current while capacitors oppose changes in voltage. Result is greatly reduced change in level between input waveform (shown for full-wave rectifiers) and output waveform. Every additional filter stage reduces content even more. Two are usually enough.

power source (or maybe at twice the ac frequency, depending upon the power supply circuits). Similarly, if it were applied to an rf power amplifier the output signal would be amplitude modulated at the ac frequency.

To prevent these unwanted effects, we use filter capacitors and inductors in the power supply dircuit. The capacitors store energy at the peaks of the pulsations, and during the valleys the stored energy goes out to the external circuit. The inductors oppose any change in current. Between them, the capacitors and inductors of the filter circuits smooth off the pulsations (Fig. 6-4) to something approaching "pure dc" such as we would get from a battery.

FCC regulations require that power supplies for transmitters operating at frequencies below the UHF region be adequately filtered to prevent AM and FM.

HOW SIGNALS PROPAGATE

Any attempt to explain accurately just how radio signals are propagated from a transmitter and received at a distant point inevitably bogs down in complexity and higher math, because the blunt truth is that science does not yet know exactly how it happens, and can account for it only by many layers of nested theories—any or all of which has rather large amounts of probable error in it.

In general, though, it appears that energy can propagate from one point to another through apparently empty space whenever it is in the form of "crossed" magnetic and electric fields. In this case, both fields are always changing in intensity, but out of phase with each other, and the result is that the energy seems to move from here to there in an ever-expanding sphere at the speed of light.

This goes on in any type of conductor, and is how electricity flows down a wire. It can also be made to happen without a wire.

To do this, it's necessary to encourage the fields to "couple" from the conductor in which they are originally active, into

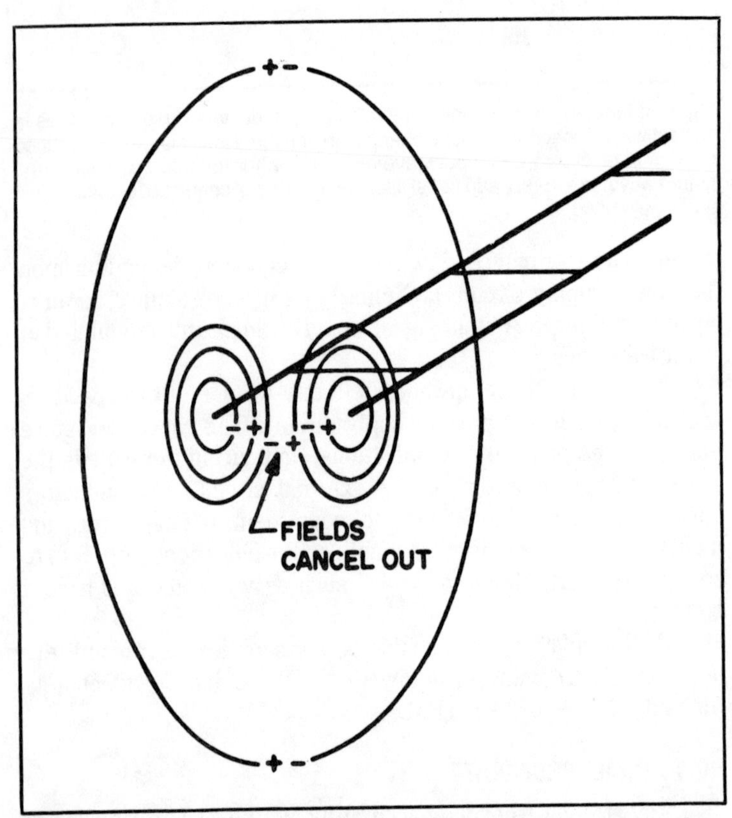

Fig. 6-5. Field cancellation is used in parallel-wire feedlines to prevent radiation. At any appreciable distance, every field radiated from one wire is canceled by equal and opposite field from the other.

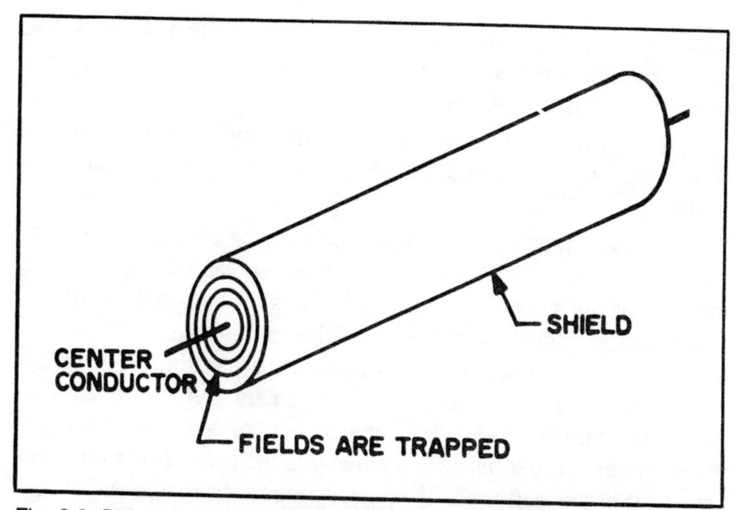

Fig. 6-6. Different principle is used in coaxial cable. Here, field from center conductor is confined by the shield, and no field is established on outside of shield. Rf is confined to the inside of the shield. Thus radiation is not a problem with coax, and it can be run almost anywhere.

"space." Fortunately, this happens automatically whenever the conductor is larger than an appreciable fraction of a wavelength.

The resulting coupling causes what we know as "radiation" of the energy. Without the effect of radiation, we could not have radio, but it's not an unmixed blessing. The trouble is that radiation happens automatically, whether we want it to or not. Unless we take careful steps to prevent it, for instance, the feedlines which connect our transmitters and our antennas will radiate just as well as the antenna. This throws a monkey wrench into our plans when we are trying to make use of directional antennas, and may even throw a high-voltage jolt through our bodies if it happens that a peak voltage point coincides with the transmitter's antenna plug!

What we do to prevent radiation is a bit complicated. We cannot actually prevent it, but we can make it cancel itself out. This is the principle behind "twin lead" feedlines (Fig. 6-5). If we feed our energy through two conductors, out of phase with each other, then all the energy radiated by one will be canceled by that from the other, and the net result will be the appearance of no radiation at all.

This is not the only possible approach. Coaxial cable (Fig. 6-6) is another. Here, the energy is confined within the shield because it has no way to escape through the outer shielding conductor. We don't care how much it tries to radiate from the inner conductor, because it cannot go anywhere. Energy always seems to follow the path of least resistance. If it finds its flow blocked by any opposition, it simply doesn't go.

When we get the energy from the transmitter up the feedline to the antenna, we have a different situation. Now we want it to radiate.

We encourage this by making the antenna exactly the proper size to support a "standing wave" at our signal frequency. This means that as each packet of energy makes its way up the feedline to the antenna, then out to the end of the antenna where it finds no place to go and so bounces back toward the feedline, and finally gets back toward the feedline, and finally gets back to the feedpoint, it finds itself in the company of a fresh packet of energy of exactly the proper phase and moving in the same direction. The net result of this activity is that energy moves from one end of the antenna to the other "in step," which keeps the actual amount of energy on the antenna wire constant at all times. None is lost by interference among the packets.

While this "standing wave" exists on the antenna, some of its energy will couple into space and be radiated. This loss of energy through radiation makes room for new energy to enter the system from the feedline, and just enough does enter to keep the standing wave on the antenna conductor at constant strength.

In practice, an antenna manages to radiate all the energy we can supply to it. The real limiting factor is the amount of power we can push up the feedline, rather than the amount which can be sustained in the standing wave.

If the antenna's length is too great, then the packets which bounce back from the ends will arrive at the feedpoint a bit late, and their phase and direction won't match that of the fresh energy arriving from the transmitter. Some cancellation will occur, and some of the energy will bounce back down toward the transmitter rather than entering the standing wave on the antenna.

Similarly, if the antenna length is too short, the packets bouncing back from the end will arrive at the feedpoint "early." Again, this will reduce the amount of energy accepted by the standing wave, and cause some of the signal to be reflected back to the transmitter.

At this point, it might appear that for any specific frequency of rf signal, only one possible antenna length would be correct. That's almost, but not quite, correct.

Actually, there isn't any way of telling two different "packets" of energy apart except for their speed and direction. This means that if an antenna is much, much too long for a given frequency of signals—so much too long, in fact, that the returning packets from the ends are delayed by one full cycle of the rf—they will appear to have bounced back from the end of an antenna of proper length.

In less theoretical terms, this boils down to the fact that almost any resonant antenna can be used at the third harmonic of the intended frequency, with almost as good results as would be had on the intended frequency. In the most practical application, it means that a 7 MHz dipole will also work well at 21 MHz.

This situation is also true at the 5th, 7th, 9th, and all other odd harmonics, but other characteristics come into play and the only practical application of this phenomenon is the use of an antenna at the third harmonic of its design frequency.

At the even harmonics (2nd, 4th, etc.) the bounce-back energy is exactly out of phase and so cancels almost completely with that coming from the transmitter. This means that most single-band antennas perform built-in rejection of even-harmonic signals. One of the surest ways to prevent trouble with even harmonics is to make use of single-band antennas.

Once the energy leaves the antenna, it still has quite a journey ahead of it before it reaches a distant receiver.

If the receiver is nearby, the process is simply the reversal of that by which it was radiated. The energy in the radiated fields "induces" current in the receiving antenna, which causes formation of a standing wave upon the antenna, and the receiver drains off some of this energy through the feedline. Since there is a standing wave, some of the received energy is reradiated, but enough remains at the receiver input to produce a "loud and clear" signal.

As the distance between transmitting antenna and receiving antenna increases, the things which happen to the signal get more complicated. While the radiating signal is expanding in a roughly spherical shape, each "ray" of the signal travels in a straight line. It does not bend around the surface of the earth, any more than a light ray does, and so communication ought to be limited to line-of-sight distances. At ultrahigh frequencies (UHF), this is true, but at lower frequencies the signal reflects from the ionosphere (Fig. 6-7) and makes possible long-distance communication.

You can think of the ionosphere as a sort of "mirror in the sky" for radio signals. When a signal hits these layers of the atmosphere, provided that conditions are right, it is reflected just as a

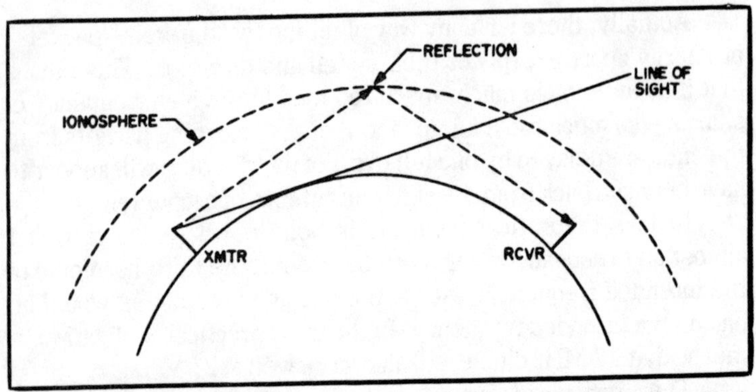

Fig. 6-7. Long-distance radio communication depends upon "mirror in sky" effect of ionized layers high above earth's surface. Signals leaving transmitting antenna reflect from the ionosphere and return to earth thousands of miles away.

mirror reflects light. The details differ slightly, because radio wavelengths are many thousands of times larger than the light wavelengths, but the principles are the same. The result is that the signal bounces down, reaching the earth's surface far beyond the horizon. This type of signal is called a sky-wave signal because it appears to arrive from the sky, or a "skip" signal because it has "skipped over" much of the earth's surface on its way up and back.

TYPICAL TRANSMITTER COMPONENTS

We've already met the basic components of any transmitter: the power supply, the frequency control device or oscillator, the power amplifier, and the modulator. Now we'll look at each in a little more detail, to see specifically how they fit together in a typical Novice station's transmitter and what function each of them performs.

As we saw earlier, the function of the power supply is to convert raw ac input power obtained from the wall outlet (in most cases) into pure dc suitable for use in powering the rest of the transmitter.

Most transmitter supplies (Fig. 6-8) contain a transformer (sometimes several, because with power amplifiers it's good practice to let the tubes reach operating temperature before applying full voltage, and it's safer to switch the 117-volt ac in the transformer primary circuit than to switch the much-higher-voltage dc in the power supply output), a rectifier, and a filter circuit.

The transformer changes the 117 volts from the wall outlet to the required value. Most vacuum tubes intended for fixed-station use require 6.3-volt ac power to heat their filaments, so at least one winding (if not a separate transformer) is used to power the filaments. The typical Novice transmitter uses from 500 to 800V dc on the plates of the tubes, so the transformer provides approximately this voltage to the rectifiers.

The rectifier circuit in a transmitter usually contains at least two and sometimes four rectifiers, although two or more of them may be contained in a single tube or solid-state "chip." Each rectifier permits current to flow through it only in one direction. The multiple rectifiers are arranged so that the power flowing in *both* directions in the ac part of the circuit is steered to the filter flowing all in the same direction. This increases efficiency of the supply.

As we saw before, the filter acts to level out the ripple left in the dc by the rectifier circuit, so that the output from the power supply is essentially pure dc and does not cause unwanted (and illegal) modulation of the output signal.

We also met the oscillator earlier, in our look at basic transmitter theory, and saw that it amounts to an amplifier which provides its own input signal, together with a frequency-controlling device.

The power amplifier, like any other amplifier, boosts power level. An rf power amplifier in a transmitter usually contains tuned circuits at both its input and its output, both to increase the efficiency of power transfer, and to cut down on the amplification of unwanted harmonics and other "garbage" which unavoidably accompanies a signal during rf amplification.

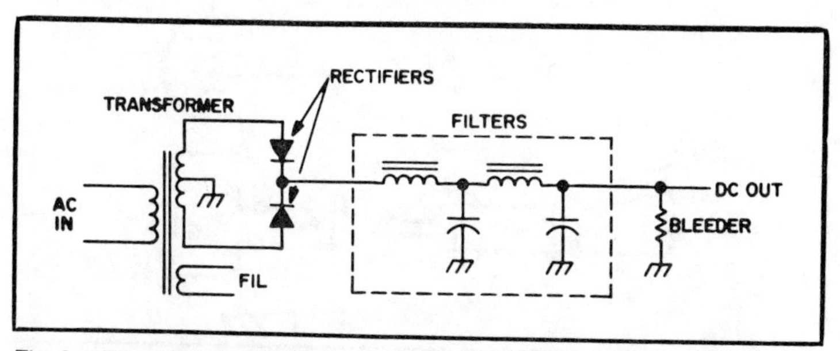

Fig. 6-8. Typical circuit for power supply consists of transformer, rectifiers, filter, and bleeder resistor. Transformer produces proper voltage from ac input power, rectifiers change it to dc, filter smooths out the dc, and bleeder provides a minimum load to improve filter action as well as serving a safety purpose.

Frequency multipliers, if any are included in the transmitter, may be part of the oscillator circuit or may be part of the power amplifier, but no well-designed transmitter will use a frequency multiplier as the final rf power amplifier. The final amplifier must operate straight through, under conditions which minimize the generation of harmonics, in order to produce a "clean" output signal.

KEYING

The modulator offers one of the widest ranges of variations in the "typical" Novice transmitter. In all cases, the "modulator" for a CW transmitter is merely the key which turns the rf on and off, but it may be located in any of several different places.

A popular arrangement (Fig. 6-9) is to place the key in the cathode lead of the final rf amplifier. This permits all other stages to operate at all times while transmitting, and usually minimizes chirp during keying. However, the strong signal produced by the transmitter circuits which are still "on" will block the receiver and prevent you from hearing any signals from outside until you turn off the oscillator in the transmitter.

To avoid this, some circuits (Fig. 6-10) key all stages. Unless carefully designed, this leads to trouble with chirp because of the change of oscillator conditions at the start of each "key down" period.

One technique sometimes used to overcome these problems is called "differential keying." In this, the key controls both the oscillator and the power amplifiers, but the oscillator is turned on a fraction of a second earlier than the amplifiers, and turned off a

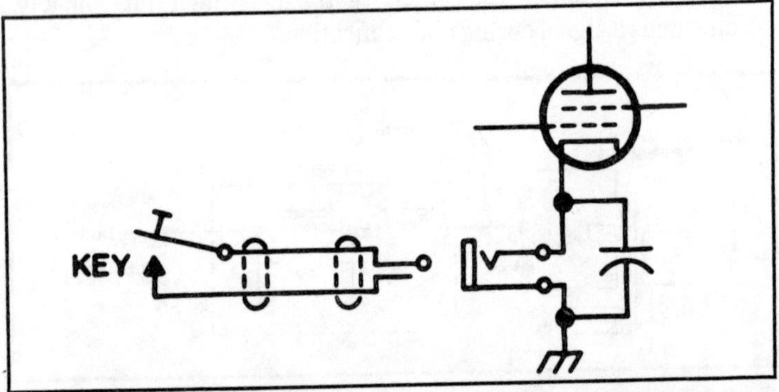

Fig. 6-9. Cathode keying of final amplifier merely makes and breaks circuit of output stage. Rest of transmitter operates, but nothing gets out to the antenna. Bypass capacitor is necessary to prevent unwanted rf feedback from key leads; rf choke may also be needed.

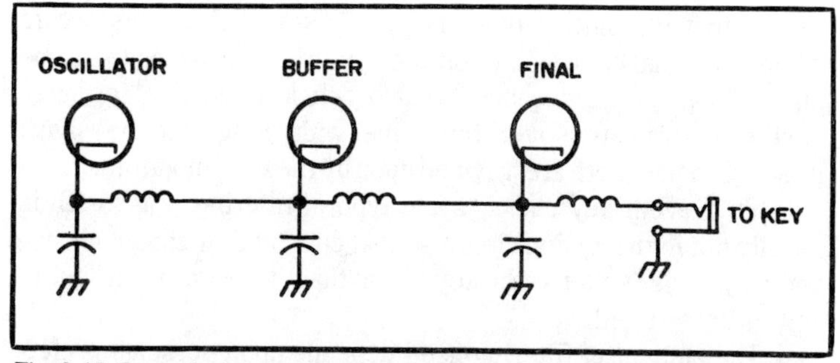

Fig. 6-10. Some designs connect the cathodes of all stages together on a single "bus," then key this bus. When key is up, no rf is present anywhere in transmitter, permitting break-in keying, but simultaneous keying often leads to excessive chirp in transmitted signal.

fraction of a second later (Fig. 6-11). The chirps caused by keying the oscillator don't get out because the amplifiers are dead when they happen. The timing is accomplished by diodes, resistors, and capacitors in the circuit, and it's a bit complex for most "first" transmitters. However, it does make it possible to achieve "full break-in" operation (in which you can hear signals from outside, between the dits and dahs of your own signal) with a clean, chirp-free signal.

Whenever the key is operated, it either makes or breaks current flow. This, in turn, means that small sparks occur. They may be so minute as to be invisible, but they're there and you can hear them on another receiver, just as you can hear a "pop" in a receiver when you turn on or off a light fixture. A similar situation,

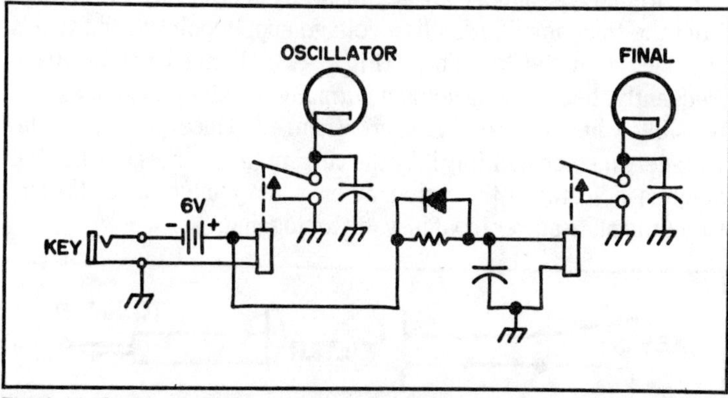

Fig. 6-11. One way to achieve "differential keying" is shown here in simplified form. Key itself operates relays. Oscillator is turned on immediately, but final remains off until capacitor across coil charges, so that chirp is gone by the time the final comes on. When key is released, reverse action occurs.

except that it's much worse, is called "key clicks;" this occurs when the signal comes on or off too abruptly, and makes the same kind of "pop". The difference is that key clicks caused by "too hard" keying are radiated over the same wide range as the signal itself—and therefore are frowned upon by the FCC monitors.

To prevent key clicks, some type of filtering (Fig. 6-12) is usually put in the keying circuit so that current flow cannot change abruptly. This "softens" the keying, so that the switch from "on" to "off" is not so rapid.

The amount of filtering used depends upon personal preferences. Many operators like a fairly "soft" filter, in the belief that it provides a signal which is easier to copy at the other end. Until the softness reaches extreme amounts, it certainly cannot make the signal more difficult to read.

PRACTICAL PROBLEMS

Now that we've accumulated some basic knowledge of transmitter theory and radio propagation, we can turn our attention to the problems which must be met in actual practice of operating a transmitter.

Naturally, its not enough to merely read the instruction sheet and turn things on. Anything as complex as even the simplest transmitter requires some operating adjustments, and FCC rules require a few measurements to be made periodically.

For one thing, you must measure the power input to the final stage of your transmitter and record it in your logbook. This measurement is almost always made by measuring the voltage from the final amplifier's plate voltage supply point to the tube's cathode, with the key down (Fig. 6-13). It need not be made frequently, because this voltage normally remains constant except possibly when the rectifiers are changed. Once you know the voltage, you then multiply it by the current (in amperes) to the final stage as indicated by the tuning meter, when you tune up on the air, to get the plate power input in watts for logging.

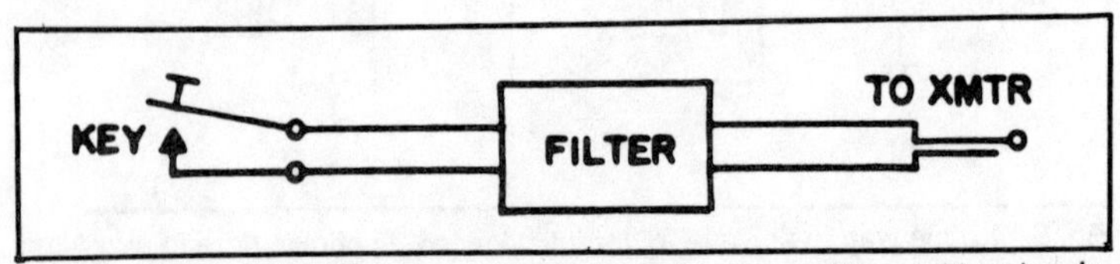

Fig. 6-12. Key-clock filter goes into circuit between key and transmitter plug. In most cases, filter should be physically as close to the key itself as you can get it. Capacitor filter should be mounted right on the key, for instance.

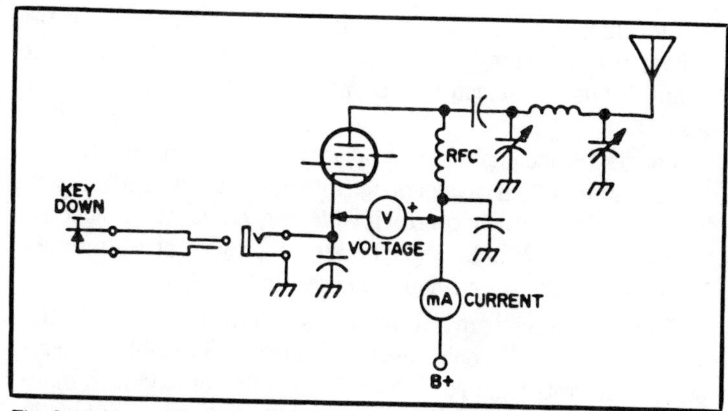

Fig. 6-13. Power input to transmitter is measured as shown here. Voltage from plate supply point to cathode, with key down, is measured. When tuneup is completed, plate current in amperes (1 mA is 1/1000 ampere).

The tuning of the transmitter is accomplished by setting the loading control to its minimum-loading position first, then applying power by closing the key, and rapidly adjusting the plate tuning control until plate current takes a sharp dip (Fig. 6-14). When the

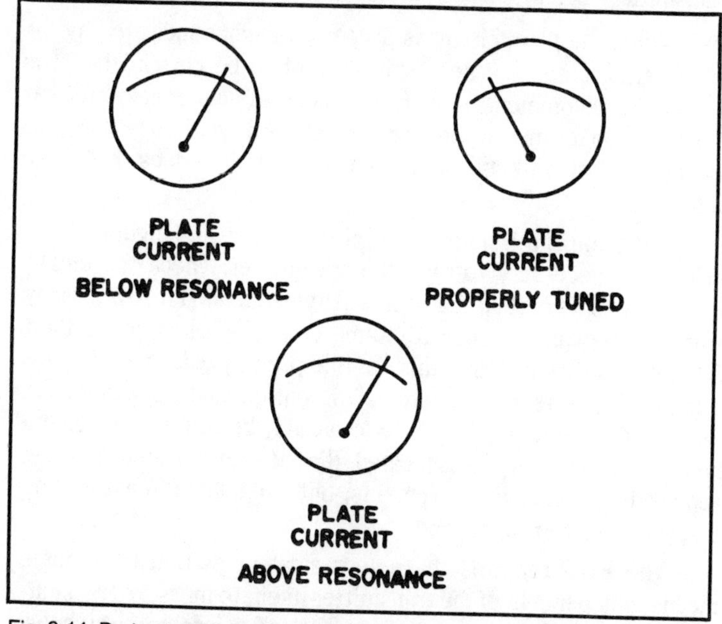

Fig. 6-14. During tuneup of a transmitter, plate current will be very high except when transmitter is tuned to resonance. At point of exact resonance, plate current is lowest. When loading, the value of the lowest plate current will rise, but dip will still be pronounced.

dip is located, tune carefully for minimum plate current. This is the point of proper tuning—but you're not yet done, because the "loading" must be adjusted to couple power from transmitter to antenna.

Loading is accomplished by adjustment of the loading control in the direction which causes plate current to increase. After each adjustment of the loading control it will be necessary to readjust the plate tuning to stay "in the dip," because the two controls interact with each other and any change in one upsets the other.

When the tuning meter indicates rated plate current, the loading is correct. If you're using a transmitter capable of more than the 250-watt legal Novice limit, then do not load it to more than 250 watts input. With lower-powered transmitters, it's best to load for maximum rf output as indicated by a power meter in the antenna feedline, because most transmitters show an increase in output power following the increase in input power only up to a certain point, and when loaded beyond that critical point, rf output-power actually drops off as the dc input power continues to increase. This not only wastes power, but may damage the transmitter itself by excessive heat dissipation.

When the plate circuit is properly tuned as indicated by the plate current meter, check grid current to be certain that it is within the recommended operating range. If your transmitter has no provision for measuring grid current, or drive, it is probably so designed that this factor is not critical and does not need adjustment.

Before putting a signal on the air, however, for longer than the brief time necessary to tune the transmitter, check the output frequency to be certain that it is what you intended it to be. Many transmitters can be tuned to 40-meter output when set to their 80-meter positions, and this is the primary cause of Novice licensees getting FCC citations for out-of-band operation. An indicating wavemeter (Fig. 6-15) can easily be built to tell you the band in which your signal is located. It's not accurate enough for the required frequency measurements, but will suffice to assure you that you're in the proper band.

The FCC requires that every amateur station have some means, independent of the transmitter itself, to measure transmitter frequency. The station receiver will suffice for this, but it helps to have a 100-kHz crystal frequency standard on hand, too, to be certain you can spot the exact band edges on the receiver dial.

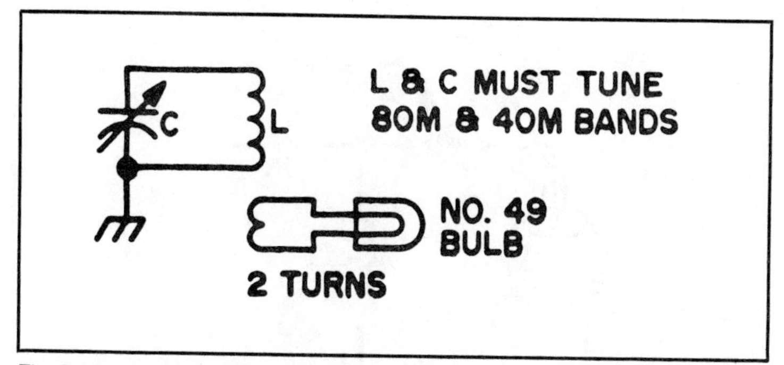

Fig. 6-15. Indicating wavemeter consists merely of tuned circuit with low-power pilot bulb coupled to it. When circuit is tuned to frequency of transmitter, bulb lights. Capacitor should be calibrated before use by reference to known frequencies.

Most receivers do not have trustworthy calibration; and an FCC "pink ticket" is the price of an error.

To measure your output frequency, simply tune to your own signal, then estimate the distance from the band edge by referring to the receiver calibration. If you use a 100-kHz standard to set up the receiver's bandspread dial, you can trust its calibration between the known 100-kHz spots. All that the FCC requires is that you be sure you are inside the legal bands. The only tricky part of this whole procedure is in making certain you are tuned to your actual signal, rather than an "image" of it (which is easy to do, because your signal is so strong in your own shack). You can usually tell the difference in strength, though, between the real signal and its image, and if you are familiar with the way your own receiver works, you can use this knowledge also to help tell them apart.

You may run into problems now and then with parasitic oscillations in the transmitter. A well-designed rig should be free of parasitics, but they sometimes develop when tubes are changed if the design is marginal. A parasitic is an uncontrolled oscillation, within an amplifier stage, which produces unwanted output and muddies up the desired output. They usually result from unexpected resonant circuits which permit the amplifier to oscillate at VHF. The normal cure for parasitics is to introduce loss into the circuit by means of "parasitic suppressors" (Fig. 6-16), which have little effect at normal operating frequencies but make the gain of the circuit too low to permit oscillation at VHF.

You may also have harmonic problems. Almost all the actual operating problems of Novices do revolve, in one way or another, around harmonics. The most common problem is that of acciden-

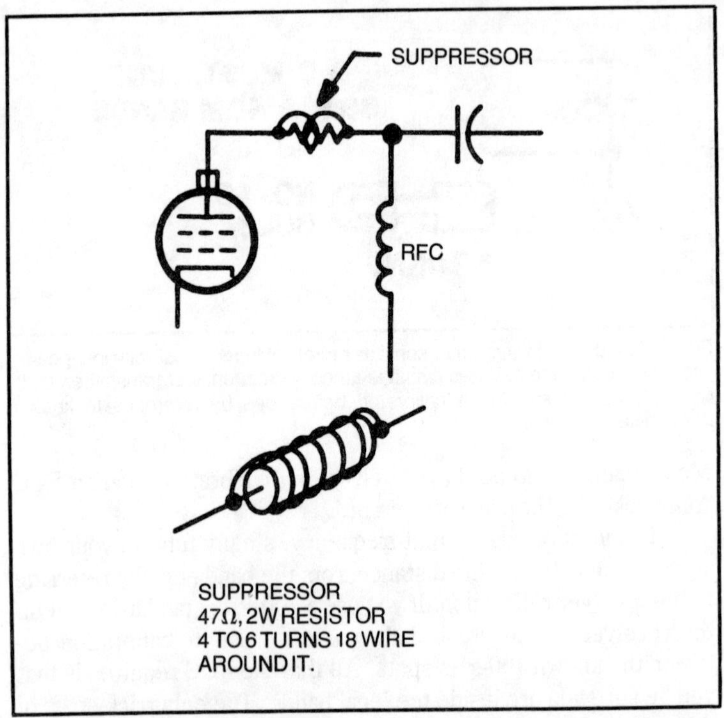

Fig. 6-16. Parasitic oscillations in rf amplifiers can be suppressed by installation of "parasitic suppressors" in plate leads as shown here. Each suppressor consists of a 47Ω 2-watt composition resistor, with 4 to 6 turns of 18-guage wire wrapped around it and soldered at each end. Result is an rf choke for high frequencies, and no effect at all at normal operating frequencies.

tally tuning up the transmitter for output at the harmonic rather than within the band, as already mentioned. This can be cured by being careful when tuning up, and always checking to be certain adjustments are right.

Use of a single-band antenna will go a long way toward curing harmonic problems, because the antenna will act like a filter in case you accidentally tune up on the wrong frequency and the resulting inability to load up the transmitter will alert you that something is not right. In addition, any even harmonics which do manage to make it through all the transmitter circuits will find it difficult to couple into space and be radiated from a single-band antenna.

If space limitations force you into using a multiband antenna rather than having separate single-band skywires for each band you intend to use, then by all means put an "antenna tuner" or "matchbox" (Fig. 6-17) between the transmitter and the antenna.

This will not only help you achieve a better match between the antenna and the transmitter, but will aid in rejecting unwanted harmonics just as the single-band antenna does.

One area in which many Novice transmitters suffer from problems which are not sufficiently obvious to be recognized by the inexperienced operators as being problems is in their modulation circuitry—or to be more specific, the keying. We've already mentioned the "chirp" problem caused by keying an oscillator, and suggested several possible cures. We've also mentioned "key clicks," but without being very specific about cures in that case.

The normal "key click filter" is simply a low-pass LC filter, which looks very much like those in your transmitter power supply. Its purpose is to round off the abrupt changes of voltage which occur when the key opens and closes, and substitute gradual changes. We show several circuits (Fig. 6-18) and you can take your pick from them, but almost any Novice transmitter can stand the installation of a key-click filter.

While, as we just said, almost any station can use one, the key-click filter's installation is a must whenever the keying of a CW transmitter causes interference to stations operating on other frequencies. Such interference is always illegal, and in addition shows the offending operator up as a person who has little respect for the rights of others—an attitude which fails to endear one to the ham ranks.

While we don't pretend that this brief discussion will make you an "instant expert" on radio transmitter theory and operation, it should give you more than adequate background in the subject to permit you to pass the Novice examination questions. Hopefully, it

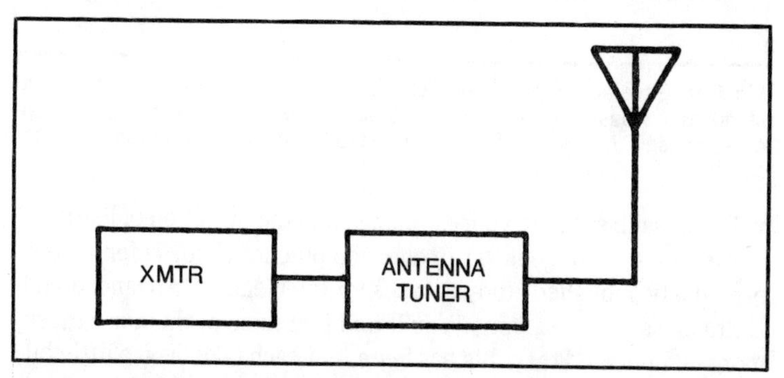

Fig. 6-17. Antenna tuner connected in feedline between transmitter and antenna helps reduce radiation of unwanted harmonics, by rejecting all signals except those to which it is tuned.

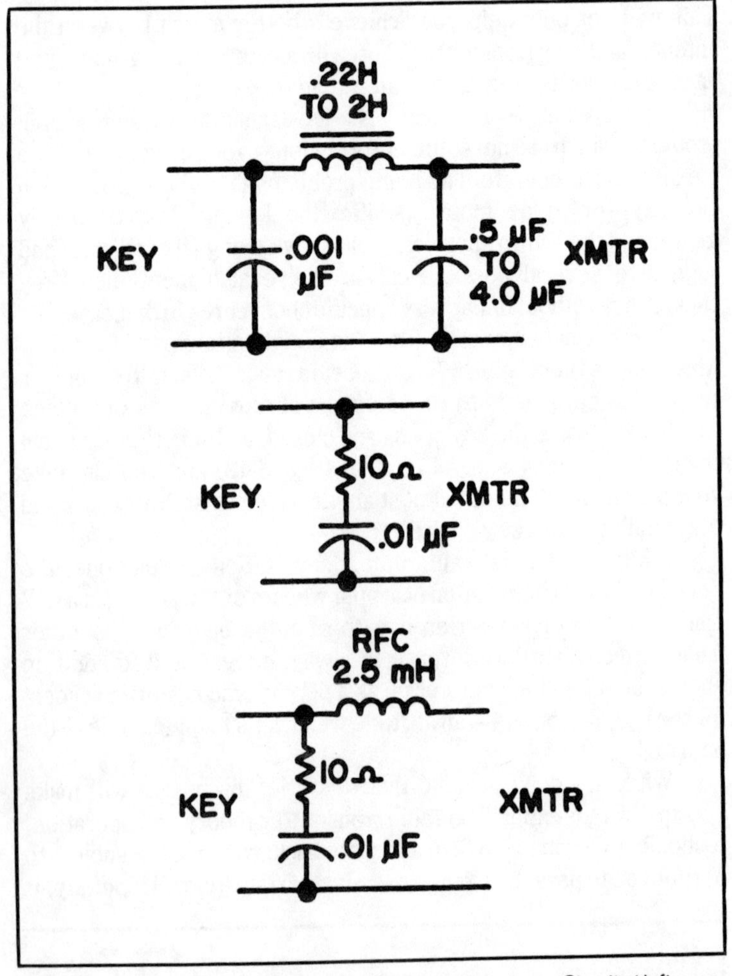

Fig. 6-18. Several circuits for key-click filters are shown here. Circuit at left uses standard low-pass filter arrangement to shape keying. Those at right reduce clicks caused by sparks as key is operated but do little toward shaping the signal.

will also have aroused your enthusiasm in the direction of learning more about what goes on in the equipment. If so, refer to our previously published study courses for the General, Advanced, and Extra class license exams. You'll find that none of them are much more complicated than this has been, but each takes you a little bit deeper into the sometimes still unexplored jungle of electronics theory—and while you're at it, you'll be working your way up the license ladder.

Chapter 7
Scope of the Exam

The FCC furnishes a study guide to help prospective amateurs prepare for their Novice exam. The study guide, which is presented here, consists of a wide range of subject areas, some of which will not be found in the actual examination. We've put in a great deal more information than you'll ever really need to pass the Novice exam, which is in itself actually quite simple.

You'll find the information useful as reference material after you've successfully passed the exam. It will speed the learning process for you and will serve as a kind of mini encyclopedia of radio. You'll learn more from your own operating experience than you possibly could from any book alone, without help—and you'll find that the more you learn in your day-to-day operating, the easier the material in this section will be for you to digest.

If you decide, after receiving your Novice license, to move up to a higher grade of license, you'll find that most of the General class exam questions are based on information contained within these pages!

A. RULES AND REGULATIONS

A.1. Basis And Purpose

The rules and regulations (Part 97) are designed to provide an amateur radio service having a fundamental purpose as expressed in the following principles:

(a) Recognition and enhancement of the value of the amateur service to the public as a *voluntary noncommercial communication service, particularly with respect to providing emergency communications*.

(b) *Continuation and extension of the amateur's proven ability to contribute to the advancement of the radio art*.

(c) Encouragement and improvement of the amateur radio service through rules which provide for *advancing skills in both the communication and technical phases of the art*.

(d) *Expansion of the existing reservoir* within the amateur radio service *of trained operators, technicians, and electronics experts*.

(e) Continuation and extension of the amateur's unique ability to *enhance international good will*.

A.2. Definitions

(1) Amateur radio service
(2) Amateur radio operator
(3) Amateur radio station
(4) Control operator
(5) Station license
(6) Primary station

Amateur radio service. A radio communications service of self-training, intercommunication, and technical investigation carried on by amateur radio operators.

Amateur radio operator. A person interested in radio technique solely with a personal aim and without pecuniary interest, holding a valid Federal Communications Commission license to operate amateur radio stations.

Station license. The instrument of authorization for a radio station in the amateur radio service.

Amateur radio station. A station licensed in the amateur radio service embracing necessary apparatus at a particular location used for amateur radio communication.

Primary station. The principal amateur radio station at a specific land location shown on the station license.

Control operator. An amateur radio operator designated by the licensee of an amateur radio station to also be responsible for the emissions from that station.

Operator license. The instrument of authorization from the F.C.C. including the class of operator privileges. This definition is also included in the definition of *amateur radio license*, below.

Amateur radio license. The instrument of authorization issued by the Federal Communications Commission comprised of a station license, and in the case of a primary station, also incorporating an operator license.

Amateur radio communication. Noncommercial radio communications, by, or among amateur radio stations solely with a personal aim and without pecuniary or business interest.

Third party traffic. Amateur radio communication by or under the supervision of the control operator at an amateur radio station to another amateur radio station on behalf of anyone other than the control operator.

Emergency communication. Any amateur radio communication directly relating to the immediate safety of life of individuals or the immediate protection of property.

The FCC Rules and Regulations (Part 97) contains other definitions, and you would be wise to at least be familiar with them. The definitions given above, however, are sufficient to pass the Novice class license examination, according to the most recent FCC study-guide syllabus (see Appendix).

A.3. Novice Class Operator Privileges

(1) Frequencies
(2) Emissions
(3) Transmitter power

Novice Class. Those amateur privileges designated and limited as follows:

(1) The power input to the transmitter final amplifying stage supplying radio frequency energy to the antenna shall not exceed 250 watts, exclusive of power for heating the cathode of a vacuum tube(s).

(2) Radio telegraphy is authorized in the frequency bands 3700—3750 kHz, 7100—7150 kHz (7050—7075 kHz when the terrestrial location of the station is not within Region 2), 21,100—21,200 kHz, and 28,100—28,200 kHz, using only Type A1 emission.

A.4. Limitations

(1) License period
(2) Antenna structures

The Novice class is normally valid for a period of 5 years from the date of issuance.

An antenna for a station in the amateur radio service which exceeds the following height limitations *may not be erected or used unless notice has been filed with both the FAA on FAA Form 7460-1 and with the Commission on Form 714 or on the license application form, and prior approval by the Commission has been obtained for:*

(1) Any construction or alteration of more than 200 feet in height above ground level at its site.

(2) Any construction or alteration of greater height than an imaginary surface extending outward and upward at one of the following slopes:

(i) 100 to 1 for a horizontal distance of 20,000 feet from the nearest point of the nearest runway of each airport with at least one runway more than 3200 feet in length, excluding heliports and seaplane bases without specified boundaries, if that airport is either listed in the Airport Directory of the current Airman's Information Manual or is operated by a federal military agency.

(ii) 50 to 1 for a horizontal distance of 10,000 feet from the nearest point of the nearest runway of each airport with its longest runway no more than 3,200 feet in length, excluding heliports and seaplane bases without specified boundaries, if that airport is either listed in the Airport Directory or is operated by a federal military agency.

(iii) 25 to 1 for a horizontal distance of 5000 feet from the nearest point of the nearest landing and takeoff area of each heliport listed in the Airport Directory or operated by a federal military agency.

(3) Any construction or alteration on an airport listed in the Airport Directory of the Airman's Information Manual.

A.5. Responsibilities

(1) Station licensee
(2) Third party
(3) Control operator

The licensee of an amateur station shall be responsible for its proper operation.

Every station when in operation shall have a control operator at an authorized control point. The control operator may be the station licensee or another amateur radio operator designated by the licensee. Each control operator shall also be responsible for the proper operation of the station.

An amateur station may only be operated in the manner and to the extent permitted by the operator privileges authorized for the class of license held by the control operator, but may exceed those of the station licensee provided proper station identification procedures are performed.

The licensee of an amateur radio station may permit any third party to participate in amateur radio communication from his station, provided that a control operator is present and continuously monitors and supervises the radio communication to insure compliance with the rules.

A.6. Station Operation

(1) Station identification
(2) One-way communications
(3) Operator license availability
(4) Station license availability
(5) Station logs
(6) Frequency measurement
(7) Point Of Communication

Station Identification

An amateur station shall be identified by the transmission of its call sign at the beginning and end of each single transmission or exchange of transmissions and at intervals not to exceed 10 minutes during any single transmission or exchange of transmissions of more than 10 minutes duration. Additionally, at the end of an exchange of telegraphy (other than teleprinter) or telephony transmissions between amateur stations, the call sign (or the generally accepted network identifier) shall be given for the station, or for at least one of the group of stations, with which communication was established.

When an amateur station is operated as a portable or mobile station, the operator shall give the following additional identification at the end of each single transmission or exchange of transmissions:

> When identifying by telegraphy, immediately after the call sign, transmit the fraction-bar DN followed by the number of the call sign area in which the station is being operated.

When an amateur station is operated outside of the 10 call sign areas and outside of the jurisdiction of a foreign government, the operator shall give the following additional identification at the end of each single transmission or exchange of transmissions:

> When identifying by telegraphy, immediately after the call sign, transmit the fraction-bar DN followed by the designator R1, R2, or R3, to show the region (as defined by the International Radio Regulations, Geneva, 1959) in which the station is being operated.

Under conditions when the control operator is other than the station licensee, the station identification shall be the assigned call sign for that station. However, when a station is operated within the privileges of the operator's class of license but which exceeds those of the station licensee, station identification shall be made by following the station call sign with the operator's primary station call sign (i.e. WN4XYZ/W4XX).

One-Way Communications

In addition to certain permitted experimental one-way transmissions, the following kinds of one-way communications, addressed to amateur stations, are authorized and will not be construed as broadcasting: (a) emergency communications, including bonafide emergency drill practice transmittions; (b) information bulletins consisting solely of subject matter having direct interest to the amateur radio service as such; (c) roundtable discussions or net-type operations where more than two amateur stations are in communication, each station taking a turn at transmitting to other station(s) of the group; and (d) code-practice transmissions intended for persons learning or improving proficiency in the international Morse code.

Availability of Operator License

The original operator license of each operator shall be kept in the personal possession of the operator while operating an amateur station. When operating an amateur station at a fixed location, however, the license may be posted in a conspicuous place in the room occupied by the operator. The license shall be available for inspection by any authorized Government official whenever the operator is operating an amateur station and at other times upon request made by an authorized representative of the Commission, except when such license has been filed with application for modification or renewal thereof, or has been mutilated, lost or destroyed, and request has been made for a duplicate license. No recognition shall be accorded to any photocopy of an operator license; however, nothing in this section shall be construed to prohibit the photocopying for other purposes of any amateur radio operator license.

Availability of Station License

The original license of each amateur station or a photocopy thereof shall be posted in a conspicuous place in the room occupied by the licensed operator while the station is being operated at a fixed location or shall be kept in his personal possession. When the station is operated at other than a fixed location, the original station license or a photocopy thereof shall be kept in the personal possession of the station licensee (or a licensed representative) who shall be present at the station while it is being operated as a portable or mobile station. The original station license shall be available for inspection by any authorized Government official at all times while the station is being operated and at other times upon request made by an authorized representative of the Commission, except when such license has been filed with application for modification or renewal thereof, or has been mutilated, lost, or destroyed, and request has been made for a duplicate license.

Station Log Requirements

An accurate legible account of station operation shall be entered into a log for each amateur radio station. The following items shall be entered as a minimum:

(a) The call sign of the station, the signature of the station licensee, or a photocopy of the station license.

(b) The locations and dates upon which fixed operation of the station was initiated and terminated. If applicable, the location and dates upon which portable operation was initiated and terminated at each location.

Frequency Measurement And Regular Check

The licensee of an amateur station shall provide for measurement of the emitted carrier frequency or frequencies and shall establish procedure for making such measurement regularly. The measurement of the emitted carrier frequency or frequencies shall be made by means independent of the means used to control the radio frequency or frequencies generated by the transmitting apparatus and shall be of sufficient accuracy to assure operation within the amateur frequency band used.

Points of Communications

Amateur stations may communicate with:

(1) Other amateur stations.

(2) Stations in other services licensed by the Commission and with U.S. Government stations for civil defense purposes in emergencies and, on a temporary basis, for test purposes.

(3) Any station which is authorized by the Commission to communicate with amateur stations.

Amateur stations may be used for transmitting signals, or communications, or energy, to receiving apparatus for the measurement of emissions, temporary observation of transmission phenomena, radio control of remote objects, and similar experimental purposes.

A.7. Administrative Sanctions

(1) Notice of violation
(2) Restricted operation

Answers To Notices Of Violations

Any licensee receiving official notice of a violation of the terms of the Communications Act of 1934, as amended, any legislative act, Executive order, treaty to which the United States is a party, or the rules and regulations of the Federal Communications Commission, shall, within 10 days from such receipt, send a written answer direct to the office of the Commission originating the official notice: *Provided, however,* that if an answer cannot be sent or an acknowledgment made within such 10-day period by reason of illness or other unavoidable circumstances, acknowledgment and answer shall be made at the earliest practicable date with a satisfactory explanation of the delay. The answer to each notice shall be complete in itself and shall not be abbreviated by reference to other communications or answers to other notices. If the notice relates to some violation that may be due to the physical or electrical characteristics of transmitting apparatus, the answer shall state fully what steps, if any, are taken to prevent future violations, and if any new apparatus is to be installed, the date such apparatus was ordered, the name of the manufacturer, and promised date of delivery. If the notice of violation relates to some lack of attention or improper operation of the transmitter, the name of the operator in charge shall be given.

Restricted Operation

If the operation of an amateur station causes general interference to the reception of transmissions from stations operating in the domestic broadcast service when receivers of good engineering design including adequate selectivity characteristics are used to receive such transmissions and this fact is made known to the amateur station licensee, the amateur station shall not be operated during the hours from 8 p.m. to 10:30 p.m., local time, and on Sunday for the additional period from 10:30 a.m. until 1 p.m., local time, upon the frequency or frequencies used when the interference is created.

In general, such steps as may be necessary to minimize interference to stations operating in other services may be required after investigation by the Commission.

A.8. Prohibited Practices

(1) Broadcasting
(2) Unidentified communications
(3) Interference
(4) Third party traffic

Broadcasting Prohibited

An amateur station shall not be used to engage in any form of broadcasting; that is, the dissemination of radio communications intended to be received by the public directly or by the intermediary of relay stations, nor for the retransmission by automatic means of programs or signals emanating from any class of station other than amateur. The foregoing provisions shall not be construed to prohibit amateur operators from giving their consent to the rebroadcast by broadcast stations of the transmissions of their amateur stations, provided, that the transmissions of the amateur stations shall not contain any direct or indirect reference to the rebroadcast.

Unidentified Communications

No licensed radio operator shall transmit unidentified radio communications or signals.

Interference

No licensed radio operator shall willfully or maliciously interfere with or cause interference to any radio communication or signal.

Third Party Traffic

The transmission or delivery of the following amateur radiocommunication is prohibited:

(a) International third party traffic except with countries which have assented thereto (see below).

(b) Third party traffic involving material compensation, either tangible or intangible, direct or indirect, to a third party, a station licensee, a control operator, or any other person.

(c) Except for an emergency communication as defined in this part, third party traffic consisting of business communications on behalf of any party. For the purpose of this section business communication shall mean any transmission or communication the purpose of which is to facilitate the regular business or commercial affairs of any party.

In paragraph (a) above, we learned "international third-party traffic except with countries which have assented thereto" is prohibited. Most countries of the world DO NOT allow amateur third-party traffic, and the U.S. amateur operator is in violation of the regulations if such communications is performed. Many of the countries which allow third-party traffic are in South America. The American Radio Relay League, a national organization of radio amateurs, usually maintains a list of those countries which allow third-party traffic. It might be wise to keep the current copy of the list handy in case you want to engage in such communications. The address is: ARRL, 225 Main Street, Newington, CT 06111.

Obscenity, Indecency and Profanity

No licensed radio operator or other person shall transmit communications containing obscene, indecent, profane words, language, or meaning.

False Signals

No licensed operator shall transmit false or deceptive signals or communications by radio, or any call letter or signal which has not been assigned by proper authority to the radio stations that he/she is operating.

A large number of "cute" practices are covered under the prohibition of "false signals." This regulation prohibits the practices of so-called "HFers" who flaunt the law by transmitting using amateur equipment on frequencies for which they are not licensed (which violates certain other rules and regulations as well). The violation of this rule is in the use of self-assigned call signs, "handles," or calls "issued" by spurious organizations such as "HF clubs." The rule also applies to amateurs who think it is cute to transmit "SOS" or "MAYDAY" signals (the internationally recognized distress signals) or the unofficial "QRRR" signal used by amateurs for distress signaling. The rule also prohibits the practice of pretending to be DX stations by signing rare or exotic false call signs for the purpose of creating a pile-up of operators who believe themselves to be in communication with a foreign country, island, or territory. False signals are always illegal, sometimes annoying, and, unfortunately, sometimes expensive and dangerous.

Communications for Hire

(a) An amateur station shall not be used to transmit or receive messages for hire, nor for communication for material compensation, direct or indirect, paid or promised.

(b) Control operators of a club station may be compensated when the club station is operated primarily for the purpose of conducting amateur radiocommunication to provide telegraphy practice transmissions intended for persons learning or improving proficiency in the international Morse code, or to disseminate information bulletins consisting solely of subject matter having direct interest to the Amateur Radio Service provided:

(1) The station conducts telegraphy practice and bulletin transmission for at least 40 hours per week.

(2) The station schedules operations on all allocated medium and high frequency amateur bands using reasonable measures to maximize coverage.

(3) The schedule of normal operating times and frequencies is published at least 30 days in advance of the actual transmissions.

Control operators may accept compensation only for such periods of time during which the station is transmitting telegraphy practice or bulletins. A control operator shall not accept any direct or indirect compensation for periods during which the station is transmitting material other than telegraphy practice or bulletins.

The meaning of this paragraph is that you cannot accept any form of compensation for the operation of an amateur radio station. Note that the rule says "transmit or receive messages for hire." This means that you are not allowed to play Western Union. The regulation further states that you may not receive any form of *material compensation* for your services, including money, gifts, or any form of "material" object. The compensation is forbidden whether it is direct (i.e. someone hands you $5), or indirect (i.e. the route of the compensation is not too blatant). The regulation also claims that the compensation need only be promised...you don't get off the hook simply because the buyer of your services welched on the deal.

There is one exception to the "no compensation" regulation, however, and that is the paid operator of a club station—such as the

A.R.R.L. station W1AW in Newington, Ct. - that is being operated for the purpose of disseminating bulletins of general interest to the amateur radio community, or to conduct more code practice sessions. There are, however, even some limitations on this form of compensation. For example, the station must operate at least 40 hours per week, and must maximize coverage by using all allocated amateur medium and high frequency bands. In addition, the schedule of these transmissions must be published at least 30 days in advance of the actual events. This is why *QST* magazine publishes the W1AW schedule.

The control operator of a club station such as W1AW may receive compensation only for the actual time that was spent transmitting the "legal" types of transmission as defined in Section 97.112. When the station is not transmitting code practice or bulletins, but is open to general operations, then the operator may not be compensated. Such transmissions, even if for a worthy cause, must be done on a voluntary basis by the control operator.

A.9 Licenses

(1) General eligibility

(a) Operator

(b) Station

(2) Renewal

(3) Commission modification

Operator License Eligibility

Novice class. Any citizen or national of the United States, except a person who holds, or who has held within the 12-month period prior to the date of receipt of his application, a Commission-issued amateur radio license. The Novice class license may not be concurrently held with any other class of amateur radio license.

General Eligibility For Station License

An amateur radio station license will be issued only to a licensed amateur radio operator, except that a military recreation station license may also be issued to an individual not licensed as an amateur radio operator (other than an alien or a representative of an alien or of a foreign government), who is in charge of a proposed military recreation station not operated by the U.S. Government but which is to be located in approved public quarters.

An amateur station license will not be issued to a school, company, corporation, association, or other organization, except that in the case of a bonafide amateur radio organization or society, a station license may be issued to a licensed amateur operator, other than the holder of a Novice class license, as trustee for such society.

Commission Modification of Station License

Whenever the Commission shall determine that public interest, convenience, and necessity would be served, or any treaty ratified by the United States will be more fully complied with, by the modification of any radio station license either for a limited time, or for the duration of the term thereof, it shall issue an order for such licensee to show cause why such license should not be modified.

Such order to show cause shall contain a statement of the grounds and reasons for such proposed modification, and shall specify wherein the said license is required to be modified. It shall required the licensee against whom it is directed to appear at a place and time therein named, in no event to be less than 30 days from the date of receipt of the order, to show cause why the proposed modification should not be made and the order of modification issued.

If the licensee against whom the order to show cause is directed does not appear at the time and place provided in said order, a final order of modification shall issue forthwith.

A.10. Sample Question

Which one of the following is not a mandatory log entry:

A. *Signature of station licensee.*
B. *Call sign of station.*
C. *Location and dates of operation.*
D. *Call signs of all stations communicated with.*
E. *All of the above are mandatory.*

Points of Communication

(a) Amateur stations may communicate with:

(1) Other amateur stations, excepting those prohibited by Appendix 2.

(2) Stations in other services licensed by the Commission and with U.S. Government stations for civil defense purposes in accordance with Subpart F of this part, in emergencies and, on a temporary basis, for test purposes.

(3) Any station which is authorized by the Commission to communicate with amateur stations.

(b) Amateur stations may be used for transmitting signals, or communications, or energy, to receiving apparatus for the measurement of emissions, temporary observation of transmission phenomena, radio control of remote objects, and similar experimental purposes and for the purposes set forth in 97.91.

This regulation provides that an amateur may contact any other licensed amateur radio station, except for those prohibited in Appendix 2 of the FCC Rules and Regulations Part 97. This appendix deals with communication with amateurs of other countries whose governments forbid their amateurs from making international contacts (or forbid amateur radio entirely!).

The U.S. amateur may also contact stations "in other services," I.E., non-amateur stations, for certain operations pertaining to Civil Defense, or for temporary tests of emergency skills and preparedness. In addition, should the FCC decide to permit certain stations in other services to contact amateur stations, then this type of operation is permitted by Part 97.

Amateur radio started from a band of persons interested in experimenting with radio ("wireless") apparatus, or in observing radio phenomena. Such experimentation is still very much a part of amateur radio. In fact, many professional engineers and scientists have used amateur radio as a quick way to obtain transmitting permission for their experiments. The *Points of Communication* regulation also permits this type of operation from amateur stations (see (b) above).

The regulations prohibits amateur stations from contacting Citizens Band or commercial stations, except as provided in (2) and (3) above. It also prohibits the amateur from knowingly contacting a bootlegger or other illegal station.

Response to an Official Notice of Violation

Any licensee receiving official notice of a violation of the terms of the Communications Act of 1934, as amended, any legislative act, Executive order, treaty to which the United States is a party, or the rules and regulations of the Federal Communications Commission, shall, within 10 days from such receipt, send a written answer direct to the office of the Commission originating the official notice: *Provided, however,* that if an answer cannot be sent or an acknowledgement made within such 10-day period by reason of illness or other unavoidable circumstances, acknowledgement and answer shall be made at the earliest practicable date with a satisfactory explanation of the delay. The answer to each notice shall be complete in itself and shall not be abbreviated by reference to other communications or answers to other notices. If the notice relates to some violation that may be due to the physical or electrical characteristics of transmitting apparatus, the answer shall state fully what steps, if any, are taken to prevent future violations, and if any new apparatus is to be installed, the date such apparatus was ordered, the name of the manufacturer, and promised date of delivery. If the notice of violation relates to some lack of attention to or improper operation of the transmitter, the name of the operator in charge shall be given.

When the FCC sends you a "pink slip" (they aren't always pink, incidentally, mine was white), there are certain steps that you must take. First, you must make your reply, complete within itself, within 10 days of receipt of the notice. The only exception is when making your response within 10 days is impossible due to unavoidable circumstances. The FCC gets to decide what circumstances are "unavoidable" in each case. In general, however, they are pretty lenient on their interpretation of "unavoidable," so long as it doesn't appear that you are trying to evade the issue.

Your answer must be complete in itself. This means that you must supply all required data at the time you reply, in addition to a copy of the original Notice of Violation (thank goodness for copy machines). If you refer to other letters, or to other offical actions or documents of the FCC, then a copy must be attached. They will not go searching for those documents!

If the notice of violation was for some technical problem with your transmission, such as observation of a second harmonic emission, then you must state fully what exactly was done to prevent reoccurance of the problem (i.e. installed a low-pass filter—give model and manufacturer — and improved shielding of the transmitter).

(a) An amateur station shall be identified by the transmission of its call sign at the beginning and end of each single transmission or exchange of transmissions and at intervals not to exceed 10 minutes during any single transmission or exchange of transmissions of more than 10 minutes duration. Additionally, at the end of an exchange of telegraphy (other than teleprinter) or telephony transmissions between amateur stations, the call sign (or the generally accepted network identifier) shall be given for the station, or for at least one of the group of stations, with which communications was established.

(b) Under conditions when the control operator is other than the station licensee, the station identification shall be the assigned call sign for that station. However, when a station is operated within the privileges of the operator's class of license but which exceeds those of the station licensee, station identification shall be made by following the station call sign with the operator's primary station call sign (i.e. WN3XYZ/W4XX).

(c) An amateur radio station in repeater operation or a station in auxiliary operation used to relay automatically the signals of other stations in a system of stations shall be identified by radiotelephony or radiotelegraphy at a level of modulation sufficient to be intelligible through the repeated transmission at intervals not to exceed ten minutes.

(d) When an amateur radio station is in repeater or auxiliary operation, the following additional identifying information shall be transmitted:

(1) When identifying by radiotelephony, a station in repeater operation shall transmit the word "repeater" at the end of the station call sign. When identifying by radiotelegraphy, a station in repeater operation shall transmit the fraction bar $\overline{\text{DN}}$ followed by the letters "RPT" or "R" at the end of the station call sign. (The requirements of this subparagraph do not apply to stations having call signs prefixed by the letters "WR".)

(2) When identifying by radiotelephony, a station in auxiliary operation shall transmit the word "auxiliary" at the end of the station call sign. When identifying by radiotelegraphy, a station in auxiliary operation shall transmit the fraction bar. $\overline{\text{DN}}$ followed by the letters "AUX" or "A" at the end of the station call sign.

(e) A station in auxiliary operation may be identified by the call sign of its associated station.

(f) When operating under the authority of an Interim Amateur Permit with privilges authorized by the Permit, but which exceed the privileges of the licensee's permanent operator license, the station must be identified in the following manner:

(1) On radiotelephony, by the transmission of the station call sign, followed by the word "interim", followed by the special identifier shown on the Interim Permit;

(2) On radiotelegraphy, by the transmission of the station call sign, followed by the fraction bar $\overline{\text{DN}}$, followed by the special identifier shown on the interim permit.

(g) The identification required by this section shall be given on each frequency being utilized for transmission and shall be transmitted either by telegraphy using the international Morse code, or by telephony, using the English language. If the identification required by this section is made by an automatic device used only for identification by telegraphy, the code speed shall not exceed 20 words per minute. The Commission encourages the use of a nationally or internationally recognized standard phonetic alphabet as an aid for correct telephone identification.

There are specific times and ways to identify an amateur radio station. Although some of the requirements (notably mobile operation) have changed somewhat over the past few years, the basics of proper identification remain the same as over the past 50 years or so. Some of our procedures are little more than common-sense methods that date back to the early days of wireless telegraphy. The regulation as shown above contains far more than the Novice must know. For the Novice-class operator, only those portions that pertain to radiotelegraph (CW) operation apply.

The first requirement is that you identify at the beginning and end of each transmission (or exchange of transmissions in the case of a "roundtable" or "ragchew"). When you are engaged in back and forth conversation (yes, Morse code telegraphy *is* a form of conversation), then it is necessary that you identify yourself not less than once every ten (10) minutes. Some amateur radio station accessories, often the so-called station console, contain a 10 minute timer that alerts the operator when the 10 minutes mandatory ID point is reached.

You are required to give the call sign of the station that you are working in addition to your own callsign when you identify. If, in the case of a network or roundtable, there are several stations, then give the call sign of the control station or at least one member of the roundtable.

The proper form for a CW ID is the callsign of the station being called or worked, followed by the letters "DE," and then your own callsign. For example: *K3RXK DE K4IPV* means that K4IPV (the author of this section) is calling K3RXK (some other guy). If K3RXK is calling K4IPV, then the order is reversed: K4IPV DE K3RXK.

If an amateur with a higher class of operator privileges operates your amateur radio station, which is a Novice class station, then certain other procedures might be necessary. As long as the operator with a higher class license stays within the Novice portions of the amateur bands, and uses power of 250 watts or less, then only your callsign is needed. For example, if K4IPV comes to your station (i.e. WN4XYZ), then it is merely necessary to sign *your* callsign: K3RXK DE WN4XYZ. Notice that K4IPV *did not* use his own call! The call of the station which you are operating is the one that you must use. Sometimes, when visiting another station or using a club station, you may hear some exotic DX stations in lower Slobovia. It sure is tempting to use your own callsign in that case in order to rack up a Lower Slobovian contact. But that is not the way you are supposed to do it!

There is one instance where the visiting operator would use his or her own callsign, but even then it is in conjunction with the callsign of the station owner. If an operator with privileges higher than the class of station being visited wants to operate outside of the band normally assigned to the station (i.e., if I visit your station and want to run down to the General or Advanced class portions of the band), then the proper procedure is to sign both callsigns separated by a slash bar (DN), with the primary station callsign first. For example, when K4IVP visits WN4XYZ and operates outside of the Novice bands, then the proper ID would be K3RXK DE WN4XYZ/K4IPV.

Okay, let's assume the best. Your studies are successful, and you pass the General class test before your Novice term is expired (a friend of mine passed the General class license while waiting for his Novice license to come in the mail!). Since you already have a call sign, and the FCC knows that your higher class license will be soon arriving, they will issue you an interim permit at the field office where you passed the examination. They will issue a special identifier that is to be used as part of your callsign until the new license arrives. This identifier is transmitted with your callsign, but is separated from it by the fraction bar (DN, or dah-di-di-dah-dit). The identifiers being issued as this written tend to be two letter combinations, such as WN4XYZ/HW.

Amateur Radio Call Signs

Callsigns identify the amateur radio station and must be transmitted as part of the station ID. Only the FCC has the authority to issue a callsign to persons or stations in the U.S. or its territories. Some CB "HF clubs" issue "callsigns" for their members, who are illegal radio bootleggers, but these callsigns are not legal and their use on the air can result in severe penalties.

Amateur callsigns are issued in a systematic manner by the FCC (97.51). The block of letter combinations used are granted to the U.S. by treaty with the other countries of the International Telecommunications Union. In "the old days," the U.S. station's callsigns would begin with the letters W, K, or N (with N being reserved for Navy and Coast Guard stations such as NAA and NSS). Today, there is a lot of confusion because of the new call-sign sequences. We now hear amateurs with N and A prefixes, and some two letter prefixes that would put some old timers in mind of exotic U.S. territorial DX stations (KS4A sounds like a Swan Island station!).

Under the old fee system, the FCC would grant a special callsign if it was legal (i.e., one of the ordinary block), not obscene, and *available* (not already assigned to somebody else). But today, the regulation is to grant callsigns only on an orderly basis as they come out of the computer.

The FCC policy regarding amateur radio call signs changes from time to time. Changes, between printings of the Rules and Regulations (Part 97), are issued to the public in the form of *public announcements*.

B. RADIO PHENOMENA

B.1. Definitions

(1) Sky wave
(2) Ground waves
(3) Surface wave
(4) Ionosphere
(5) D,E,F_1 and F_2 layers
(6) Skip zone
(7) Skip distance
(8) Wavelength

Sky Waves & Ground Waves

When a radio wave leaves a vertical antenna, the field pattern of the wave resembles a huge donut lying on the ground with the antenna in the hole at the center, as seen in Fig. 7-1. Part of the wave moves *outward in contact* with the ground to form the *ground wave*, and the rest of the wave moves upward and outward to form the *sky wave*, as shown in Fig. 7-2. The ground and sky portions of the radio wave are responsible for two different methods of carrying the messages from transmitter to receiver. The ground wave is used both for short-range communications at high frequencies with low power and for long-range communication at low frequencies and with very high power.

The sky wave is used for long range high frequency communication.

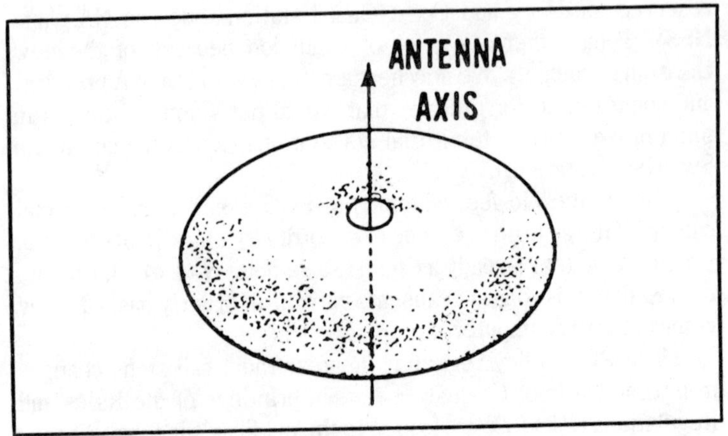

Fig. 7-1. Vertical antenna field pattern.

As the ground wave passes over and through the ground, this wave induces a voltage in the earth, setting up eddy currents. The energy used to establish these currents is absorbed from the ground wave, thereby weakening it as it moves away from the transmitting antenna. Increasing the frequency rapidly increases the attenuation so that ground wave transmission is limited to relatively low frequencies.

Since the electrical properties of the earth along which the *surface wave* travels are relatively constant, the signal strength from a given station at a given point is nearly constant. This holds essentially true in all localities except those having distinct rainy and dry seasons. There the difference in the amount of moisture causes the conductivity of the soil to change.

That portion of the radio wave which moves upward and outward is not in contact with the ground and is called *sky wave*. It behaves similarly to the ground wave. Some of the energy of the sky wave is refracted (bent) by the ionsphere so that it comes back toward the earth. Some energy is lost in dissipation to particles of the atmospheric layers. A receiver located in the vicinity of the returning sky wave will receive strong signals even though several hundred miles beyond the range of the ground wave.

Ionosphere

The ionosphere is found in the rarefied atmosphere approximately 40 to 50 miles above the earth. It differs from other atmospheric parts in that it contains a much higher number of positive and negative ions. The negative ions are believed to be atoms whose energy levels have been raised to a high level by solar bombardment of ultraviolet and particle radiations. The rotation of the earth on its axis, the annual course of the earth around the sun, and the development of sunspots all affect the number of ions present in the ionosphere, and these in turn affect the quality and distance of electronic transmissions.

The ionosphere is constantly changing. Some of the ions are returning to their normal energy level, while other atoms are being raised to a higher energy level. The rate of variation between high and low level of energy depends upon the amount of air present and the strength of radiation from the sun, as well as the propagation with relation to the earth's magnetic field.

At altitudes above 350 miles, the particles of air are far too sparse to permit large-scale energy transfer. Below about 40 miles altitude, only a few ions are present because the rate of return to a normal energy level is high. Ultraviolet radiations from the sun are absorbed in passage through the upper layers of the ionosphere so that below an elevation of 40 miles, too few ions exist to affect, materially, sky wave communication.

Densities of ionization at different heights make the ionosphere appear to have layers. Actually, there is thought to be no sharp dividing line between layers, but for the purpose of discussion, such a demarcation is indicated.

The ionized atmosphere at an altitude of between 40 and 50 miles is called the *D* layer. Its ionization is low and it has little effect on the propagation of radio waves except for the absorption

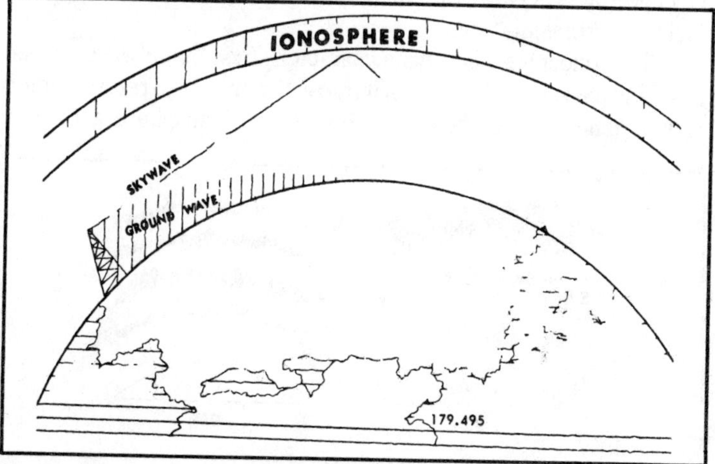

Fig. 7-2. Ground and sky waves.

of energy from the electronic waves as they pass through it. The *D* layer is present only during the day. Its presence greatly reduces the field intensity of transmissions that must pass through daylight zones.

The band of atmosphere at altitudes between 50 and 90 miles contains the so called *E* layer. It is a well defined band with greatest density at an altitude of about 70 miles. This layer is strongest during daylight hours, and is also present, but much weaker, at night. The maximum density of the *E* layer appears at about noon local time. The ionization of the *E* layer at the middle of the day is sometimes sufficiently intense to refract frequencies up to 20 MHz back to the earth. This action is of great importance to daylight transmissions for distances up to 1500 miles.

The *F* layer extends approximately from the upper limits of the ionosphere. At night only one *F* layer is present, but during the day, especially when the sun is high, this layer often separates into two parts, F_1 and F_2, as shown in Fig. 7-3. As a rule, the F_2 layer is at its greatest density during early afternoon hours, but there are many notable exceptions of maximum F_2 density existing several hours later. Shortly after sunset, the F_1 and F_2 layers recombine into a single F layer.

In addition to the layers of ionized atmosphere that appear regularly, erratic patches of ionized atmosphere occur at *E* layer heights in the manner that clouds appear in the sky. These patches are referred to as *sporadic-E* ionizations.

Sometimes sporadic ionizations appear in considerable strength at varying altitudes and actually prove harmful to electronic transmissions.

The ionosphere has many characteristics. Some waves penetrate and pass entirely through it into space, never to return. Other waves penetrate but bend. Generally, the ionosphere acts as a

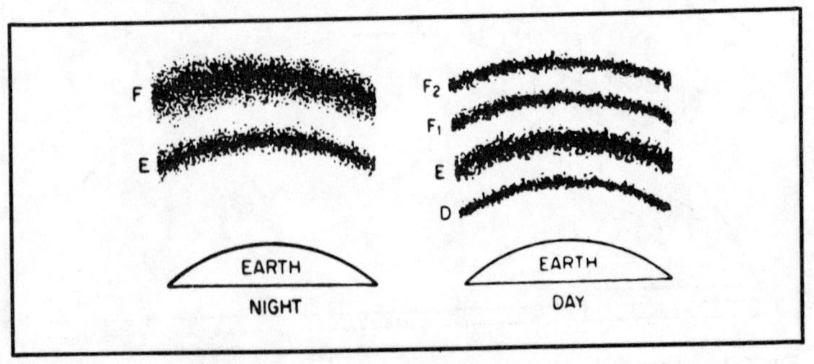

Fig. 7-3. E and F layers of the ionosphere.

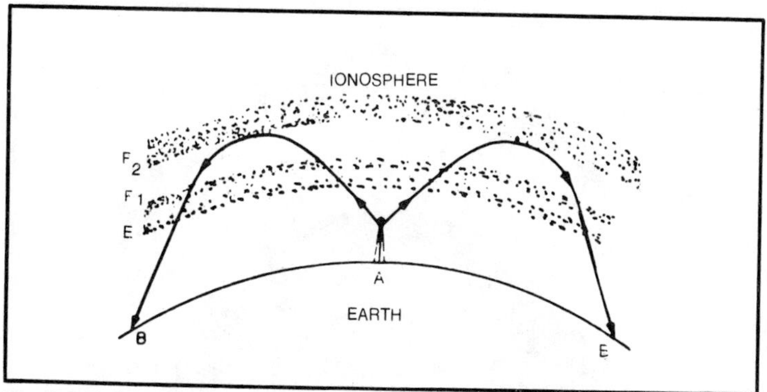

Fig. 7-4. Refraction of the sky waves by the ionosphere.

conductor, and absorbs energy in varying amounts from the electronic wave. The ionosphere also acts as an electronic mirror and refracts the sky wave back to the earth, as illustrated in Fig. 7-4. Here the ionosphere does by refraction what water does to a beam of light.

The ability of the ionosphere to return an electromagnetic wave to the earth depends upon the angle at which the sky wave strikes the ionosphere, the frequency of the transmissions, and ion density. When the wave from an antenna strikes the ionosphere at an angle, the wave begins to bend. If the frequency and angle are correct and the ionosphere is sufficiently dense, the wave will eventually emerge from the ionosphere and return to the earth. If a receiver is located at either of the points B, Fig. 7-4, the transmission from point A will be received.

The sky wave in Fig. 7-5 is assumed to be composed of rays that emanate from the antenna in three distinct groups that are identified according to the angle of elevation. The angle at which the group 1 rays strike the ionosphere is too nearly vertical for the rays to be returned to earth. The rays are bent out of line, but pass completely through the ionosphere and are lost.

Skip Zone and Skip Distance

Between the point where the ground wave is completely dissipated and the point where the first sky wave returns, no signals will be heard. This area is called the *skip zone* and is illustrated in Fig. 7-6. The skip zone for the lower high frequencies (3 to 8 MHz) will be greater at night than during the day. As a general rule, it can be said that as the frequency decreases, the *skip distance* decreases.

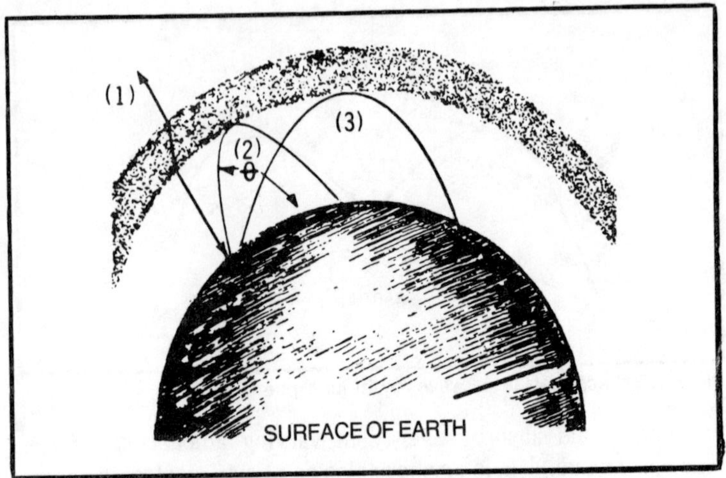

Fig. 7-5. Effect of the angle of departure on the area of reception.

Wavelength

A wavelength is the *physical size of a single cycle of a transmitted radio wave*. At a frequency of 7.1 MHz (7100 kHz), precisely 7.1 million individual cyclic waves are transmitted from an antenna during each second of operation. To determine the physical length of each wave, all you have to know is the speed at which the waves propagate—about 300 million meters per second. Wavelength λ is equal to speed of light c divided by frequency f, or $\lambda = c/f$. The wavelength of a 7.1 MHz signal, then, is

300 (millions of meters per second)
7.1 (millions of hertz, or cycles per second)
= 42.25 meters
The wavelength is 42.25 meters.

B.2. Wave Characteristics

(1) Polarization
(2) Speed versus medium through which waves travel
(3) Spreading
(4) Radiation fields

Wave Characteristics

The radiated energy from an antenna is in the form of a spreading sphere. A small section of this sphere is called a *wavefront*; it is perpendicular to the direction of travel of the energy. All energy on this surface is in phase. Usually all points on the

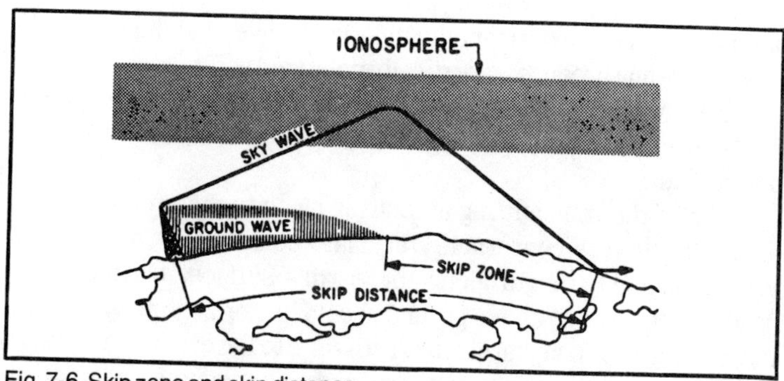

Fig. 7-6. Skip zone and skip distance.

wavefront are at equal distances from the antenna. Because of *spreading*, the farther from the antenna, the less spherical the wave appears. At a considerable distance the wavefront can be considered as a plane surface at right angles to the direction of propagation.

The *radiation field* is made up of magnetic and electric lines of force which are always at right angles to each other. Most electromagnetic fields in space are said to be linearly polarized. The direction of polarization is the direction of the *electric* vector. That is, if the electric lines of force are horizontal, the wave is said to be horizontally polarized (Fig. 7-7A), and if the lines are vertical, the wave is said to be vertically polarized.

As the electric field is parallel to the axis of the antenna, the antenna is in the plane of polarization. The horizontally placed

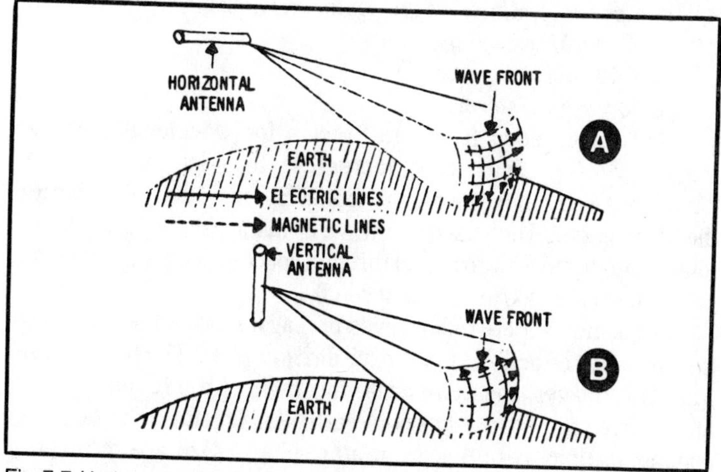

Fig. 7-7. Horizontal and vertical polarization.

antenna produces a horizontally polarized wave, and the vertically placed antenna produces a vertically polarized wave.

For maximum absorption of energy from electromagnetic fields, it is necessary that an antenna be located in the plane of polarization.

When the transmitting antenna is close to the ground insofar as propagation is concerned, vertically polarized waves cause a greater signal strength along the earth's surface. On the other hand, antennas high above ground should be horizontally polarized to get the greatest signal strength possible to the earth's surface.

When radio waves travel through some medium—a length of wire, air, an antenna—the propagation velocity is less than the speed of the same wave in space. The difference between air and space are negligible, so the equation above is applicable for calculations of propagated signals in general. When lengths of waves to be propagated along lines must be known, however, a correction factor must be incorporated into the equation. Generally, the wavelength of a signal on an antenna is about 5% shorter than the same wave in free space; and the wavelength of a signal on a transmission line (used to feed an antenna) may be more than 35% shorter than the same wave radiated into space.

B.3. Wave Propagation

(1) Wavelength versus frequency
(2) Speed of radio waves
(3) Affects of ionization upon wave propagation
(4) Atmospheric conditions versus communications
(a) Daylight versus night hours
(b) Seasonal variations
(c) Ionospheric storms

As can be seen from the formula for wavelength, as the frequency of a radio wave increases, the size of the wave diminishes—in other words, the higher the frequency the shorter the wavelength. The speed of a wave is an unchanging constant as long as the wave is propagated through space—300 million meters per second (186,000 miles per second).

Ionization affects radio waves propagated from the surface of the earth by refraction, as we have already seen. The more dense the ionized layer, the more acute the angle of refraction, and the more waves are bounced back to earth. As a result, long-distance communications resulting from atmospheric skip are most pronounced during daylight hours, when the earth is being bombarded with ions from the sun. Summer days offer the most intense ionospheric refraction, and winter nights the least. Ionospheric storms—sunspots and solar flares—tend to occur in 11-year cycles. During periods of peak sunspot activity, multiple-hop communications is considerably enhanced, and amateurs are able to communicate over great distances on high frequencies.

B.4. Propagation Modes

(1) Determination of the optimum working frequency

(2) F_2 layer reflection and fading

At sufficiently high frequencies, regardless of the angle which the rays strike the ionosphere, the wave will not be returned to the earth. The *critical frequency* is not constant but varies from one locality to another, with the time of day, with the season of the year, and with the sunspot cycle.

Because of this variation in the critical frequency, nomograms and frequency tables are issued that predict the *maximum usable frequency* (MUF) for every hour of the day for every locality in which transmissions are made. Nomograms and frequency tables are prepared from data obtained experimentally from stations scattered all over the world. All this information is pooled, and the results are tabulated in the form of long-range predictions that remove some of the guesswork from transmissions.

The sky wave path of a signal propagated at two different vertical angles is illustrated in Fig. 7-8. When the vertical angle is Θ1, the signal is returned to the earth at point A, reflected back to the ionosphere and reappears at point B. If the same signal is transmitted at a lower vertical radiation angle Θ2, it can reach point B in a single hop. The signal transmitted at angle Θ1 will

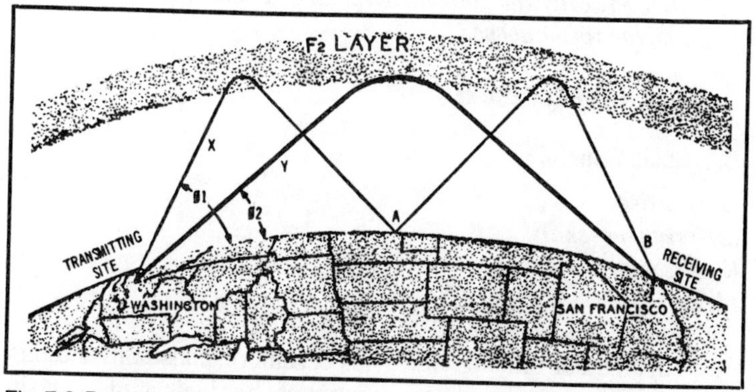

Fig. 7-8. Path of a sky-wave signal propagated at two different vertical angles.

suffer more ionospheric and ground absorption losses than that signal transmitted at angle Θ2. In general, single-hop transmission result in greater field intensities at a distant point than multiple-hop transmissions. By inspection of Fig. B8, it is evident that longer distances can be covered by multiple-hop transmissions as the vertical radiation angle is decreased. There is a limit to the improvement obtained by low angle radiation, because absorption and other factors make operation on vertical radiation angles below 3 degrees impractical.

When a received signal varies in intensity over a relatively short period of time, the effect is known as fading, which is one of the most troublesome problems encountered in electronic reception.

There are several conditions which can produce fading. One type of fading is prevalent in areas where sky waves are relied upon for transmission. Figure 7-9 shows two sky waves traveling paths of different lengths, thereby varying about the same point out of phase and thus producing a weakening of the signal. For instance, if a portion of the transmitted wavefront arrived at a distant point via the E layer and another via the F_2 layer, a complete cancellation of signal voltages would occur if the waves arrived 180° out of phase and with equal amplitude. Usually, one signal is weaker than the other; therefore, a usable signal is obtained.

B.5. Sample Question

A radiation field contains

A. only an electric component.
B. an electric and magnetic component.
C. only a magnetic component.
D. a vertically and horizontally polarized component.
E. None of the above.

C. OPERATING PROCEDURES

C.1. Basic Principles

(1) Courtesy
(2) Frequency sharing
(3) Frequency selection
(4) Zero-beating a signal
(5) Choice of telegraphy speed

An amateur radio operator has certain responsibilities to his peers. While courtesy per se cannot be legislated, certain

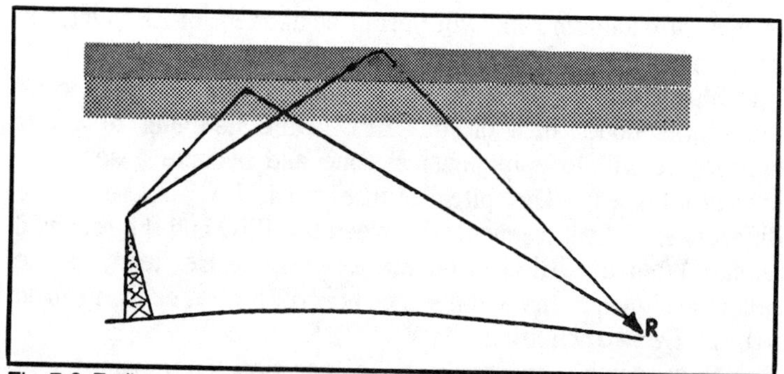

Fig. 7-9. Fading caused by arrival of two sky waves at the same point (R) out of phase.

operating practices can be and in fact are required. If you follow the Golden Rule in your station operation, you won't go wrong.

Before placing your transmitter on the air, you are required to listen first on the frequency you intend to use. If another station is using that frequency, you are required to wait until the frequency is clear or, alternatively, change frequency.

Always remember that you have no exclusive claim to any frequency. You are required to share the amateur bands you use with others operating under the same license class. Thus, you should restrict your calls to no more time than necessary to complete your communication. If you're making a blind general call (CQ), make but a few initial calls and then listen and tune around that frequency for a response.

Zero-beating a signal is necessary to ensure that your transmitter is on the same frequency as the station that you are trying to call. Otherwise, the operator of the other station won't even hear you, much less respond. What does the term *zero-beat* mean? Perhaps the best way to explain it is in the form of an analogy. Go to a piano and strike *A* above *middle C*. Note the tone. Now, strike *middle C*. Note that the tone was different. Now simultaneously strike both *middle-C* and *A-above-middle-C*. The result will be a complex tone consisting of the two *original* tones plus their *sum* (A+C) and *difference* (A– C). It is the difference tone that we are going to use to zero beat a station, and this tone (newly created by the process of beating two signals together) is called a *beat note*.

In the communications receiver we will have a circuit called a *beat frequency-oscillator* (often as part of another circuit called a product detector), or BFO. The BFO is sometimes a variable frequency oscillator, and will permit adjustment of its frequency back and forth across the receiver's *intermediate frequency* (IF). The BFO produces a beat note that allows us to more easily copy the morse code (turn off the BFO while listening to a CW station...it will lose its musical tone and become a series of staccato hisses). The pitch of the musical CW note is the difference, or beat frequency between the BFO and the received signal. When the BFO is on the exact same frequency as the received signal, then the difference is zero, so the receiver is said to be in *zero beat* condition.

Now how does this apply to the operation of a transmitter? If we turn on the transmitter's oscillator, and listen for it in the receiver while also listening to the station that we wish to call, then the beat note heard will be the difference between the transmitter oscillator and the other station's frequency. We adjust the transmitter VFO until the beat note is zero (i.e. the transmitter is *zero beat* with the other station).

The FCC study guide asks you to know how to zero-beat a station. The material in the above paragraph tells you how to do it. But this is not always the best way to approach the problem. In many cases, experienced operators will not zero beat the transmitter with the other station. Remember, in CW we are using a BFO in the receiver that causes the pitch of the CW signal to be 300 to 1200 hertz (depending upon what the operator perceives as comfortable!). This means that the received frequency is offset by 300 to 1200 hertz from the transmitter's zero-beat frequency. A better solution, in the opinion of many, is to adjust your transmitter VFO until the beat-note from the speaker or earphones has the same pitch as the received signal. If, for example, you adjust the receiver BFO for an 800 Hz pitch from the received CW signal, then don't zero-beat the signal with your transmitter VFO...adjust the VFO for the same 800 Hz pitch. This procedure will make your transmitted frequency the same as the other guys transmitted frequency. This was not too much of a problem in "the olden days" when everybody used separate receiver/transmitter combinations. But, today, most operators use *transceivers* in which the transmitter and receiver are in a common package. Unless the other guy's transceiver is equipped with *receiver incremental tuning* (RIT) or some similar feature, it will be impossible for him to tune in your signal without also changing his own transmitter frequency. So, when he comes back to you, he will have shifted frequency.

Incidentally, on today's crowded bands, it is a pain to have to continually chase a signal. When you acquire a lot of CW skill, then you will begin to be able to copy a single station in a pile-up. Little differences in BFO pitch are one point of recognition that will help you discriminate between the station that you have been working and all of the "other guys." This is why it is important to keep the beat-note constant... you certainly don't want him to come back on a slightly different frequency than before!

The choice of CW speed is important to establishing communications. There are two considerations here, one legal and one practical. On the legal side, telegraphy speed affects the *bandwidth* of your transmitted signal. A higher speed means a wider signal. If you insist on operating close to the edge of the band, then a too-fast speed could conceivably put you into the out-of-band region. This isn't too likely at the speeds most novices can handle, but it is a possibility. The second aspect is purely practical. If you show off, and transmit a "CQ" (general call) at some blistering speed that is far greater than you can copy, then one of two things is likely to happen. One, you will scare off most of the people in the band who might be looking for contacts. Remember, most of the other occupants of the novice band are *novices* and can't copy 20 words per minute! Second, you just might have to eat crow when some old duck who has been operating the transcontinental CW traffic nets for 40 years come blazing back to you at the speed you set. Your plaintive "QRS, pse OM" (go slower, please, old man) may well be met with a disdainful "QLFF LID" (now try sending with your left foot, turkey!), assuming the old timer is impolite.

On the other side of the coin, when you have gained some speed, don't try to force other newcomers to the amateur radio bands to try copying your speed. They cannot, and you would be simply rude to expect them to. Set your CW speed at a level that is agreeable to both the other operator and yourself.

C.2. Telegraphy Procedures

(1) Q signal system
(2) RST reporting system
(3) Standard abbreviations

Operating a CW Station

The Novice class operator is limited by law to the use of radiotelegraphy emissions (A1 or CW) in which the RF carrier signal is turned on and off in the pattern of dots and dashes of the international radiotelegraph code. The code is shown for reference in Table 7-1, although teaching the code is beyond the scope of this book.

The radiotelegraph operator uses some special abbreviations and signals to make communications easier. The special signals are the so-called "international Q-signals" that are designed to aid communications between operators who don't speak each other's language. The R-S-T reporting system is a special system used to inform other operators of the quality of their transmissions. There are also a large number of abbreviations used in telegraphy. These are given below. The need for abbreviations becomes obvious when you try to transmit certain long passages, especially when certain words keep popping up over and over again. We could transmit entire words, but the abbreviations are a lot easier. For example, the word "please" in Morse code is .--. .-.. . .- ... . while the abbreviation for please ("pse") is merely .--. ... . You can, perhaps, see why operators rapidly become familiar with a host of abbreviations and other symbols.

Another class of special symbols are the operating symbols, or message symbols. These tell the other operator exactly what you are planning to do next, or what was just done. The symbol AR, for example, is used after the call to another station, but before contact has been firmly established. For example: K3RXK DE K4IPV $\overline{\text{AR}}$. The bar over the letters "AR," incidentally, indicate that they are to be transmitted as a single character (i.e., *didahdidahdit* instead of *didah didahdit*).

RST Reporting System. The RST system is used to give other operators a qualitative report of how they sound to you. This system is terribly subjective, and doesn't really provide any hard data that is of use to you in engineering a new amateur station. In some cases, the other stations may well send only three RSTs, a high, low and medium. Although there is some point to this madness, the RST reports received must all be considered with a grain of salt unless they are *uniformly low*. The RST designation means *Readibility, Strength,* and *Tone* (the three principle aspects of a radiotelegraph signal). The readibility is graded on a scale of 1 to 5 (although few amateurs ever use any digit other than 5!), strength on a scale of 1 to 9, and tone on a scale of 1 to 9. The strength scale is based very loosely on the "S-meter" found on the communications receiver. The meanings of the RST scales are:

Readability

1 - Unreadable
2 - Barely readable, some words distinguishable (radiotelephone)

3 - Readable, but with considerable difficulty
4 - Readable with practically no difficulty
5 - Perfectly readable; ideal copy

We, perhaps, should not be too harsh on amateurs who never give less than a R5. Perhaps they are capable of reading every CW signal they work, regardless of how much interference exists on today's crowded bands. If you make contact, and can make out the code characters only with the utmost concentration (because of too many signals in the receiver at the same time!), then R3 would be appropriate. Most signals are probably an R4, although most people simply give the other station an R5 report.

Strength

1 - Faint, barely perceptible, signals
2 - Very weak signals
3 - Weak signals
4 - Fair signals
5 - Fairly good signals
6 - Good signals
7 - Moderately strong signals
8 - Strong signals
9 - Extremely strong signals

The strength signal (i.e., the "S" of the RST system) is used to tell the other operator how strong his signal is at your location. Many operators tend to make the task easier by using only three levels: 5 or 6 for a weak signal, 7 for an average signal and 9 for anything from a moderately strong signal to a billion-watt block buster. These numbers appear to correspond (very) roughly with the S1 through S9 markings of the signal-strength meter on most communications receivers. Note, however, that most S meters are totally inaccurate when receiving CW.

Tone

1 - Sixty cycle ac or less, very rough and broad.
2 - Very rough ac, very harsh and broad.
3 - Rough ac tone, rectified but not filtered.
4 - Rough note, some trace of filtering.
5 - Filtered rectified ac but strongly ripple-modulated.
6 - Filtered tone, definite trace of ripple modulation.
7 - Near pure tone, trace of ripple modulation.
8 - Near perfect tone, slight trace of modulation.
9 - Perfect tone, no trace of ripple or modulation of any kind.

If the signal has the characteristic steadiness of crystal control, add the letter X to the RST report. If there is a chirp, the letter C may be added. Similarly for a click, add K.

You are not harming the other operator if you give an "optimistic" signal report with regards to R and S, but don't lie about *T*. A bad tone can mean that the other station is (probably inadvertently) sending out an illegal signal, and may wind up in trouble with the FCC. The government has some very definite ideas about technical standards, and one must obey them. If you hear some dude on the air with a raw AC buzz, or some ripple modulation, on the signal, then ***tell them*** about it so the rig can be repaired properly.

When you start working DX stations, however, you will find some of them that have some hum or ripple on their signals because they may not have technical standards as tough as those in the U.S. Give them a correct report anyway; call them the way you hear them, and let the DX station operate the way he sees fit.

Most of the operating symbols and the Q-signals that you will need to know in order to pass the Novice class examination (plus operate an amateur radiotelegraph station when the "ticket" arrives in the mail) are given in Chapter 1. You are advised to return to Chapter 1 and reread that section (unless you are already an expert).

C.3. Sample Question

The signal *QRM* generally means

A. *a transmission is experiencing interference.*
B. *a frequency is varying.*
C. *a reply is requested on a certain frequency.*
D. *the sending speed is too fast.*
E. *the previous message is to be repeated.*

D. EMISSION CHARACTERISTICS

D.1. Definitions

(1) Spurious radiation
(2) Key clicks
(3) Carrier frequency
(4) Continuous waves

Table 7-1. The Radiotelegraph Code, Also Known as International Morse Code.

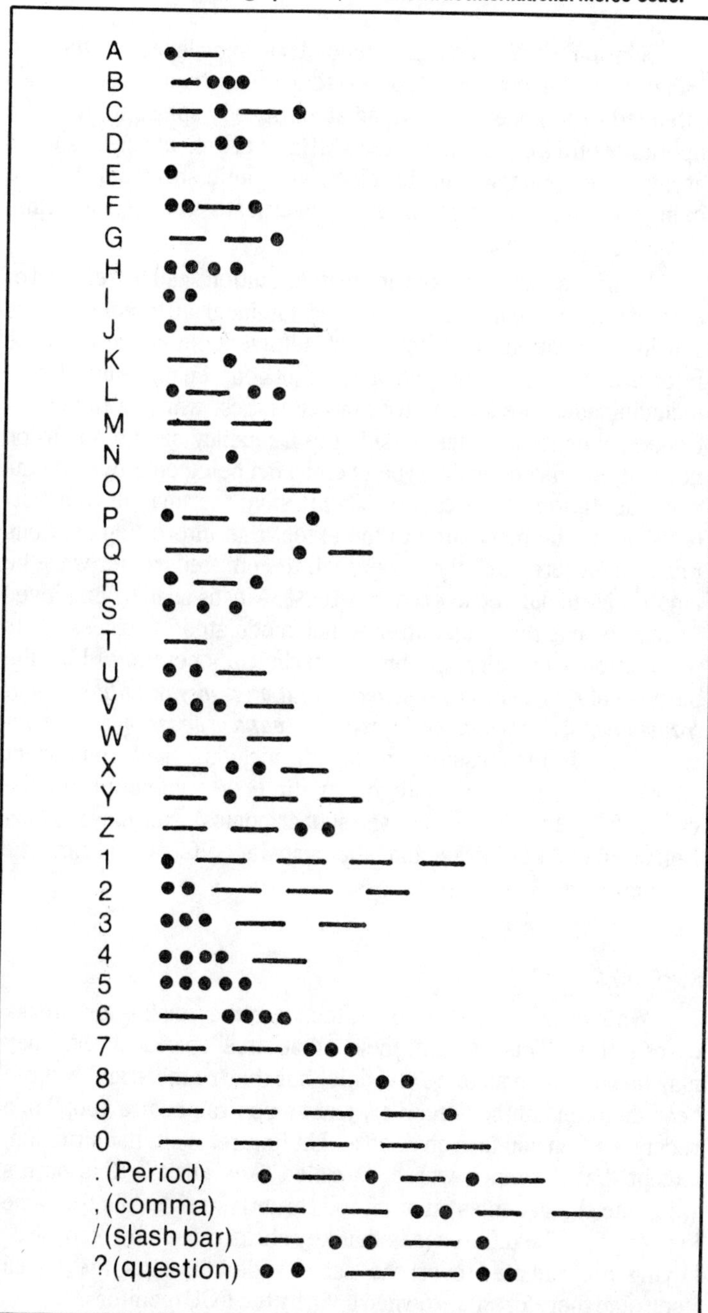

A	• —
B	— • • •
C	— • — •
D	— • •
E	•
F	• • — •
G	— — •
H	• • • •
I	• •
J	• — — —
K	— • —
L	• — • •
M	— —
N	— •
O	— — —
P	• — — •
Q	— — • —
R	• — •
S	• • •
T	—
U	• • —
V	• • • —
W	• — —
X	— • • —
Y	— • — —
Z	— — • •
1	• — — — —
2	• • — — —
3	• • • — —
4	• • • • —
5	• • • • •
6	— • • • •
7	— — • • •
8	— — — • •
9	— — — — •
0	— — — — —
. (Period)	• — • — • —
, (comma)	— — • • — —
/ (slash bar)	— • • — •
? (question)	• • — — • •

Spurious Emissions

Spurious emissions are defined as signals generated and radiated by an equipment or system other than those signals intended to be generated and radiated. Strictly speaking, if you're operating on a frequency of 7.1220 MHz, there should be no signals at any other point within the electromagnetic spectrum that are being radiated by your equipment, power leads, transmission line, and antenna system.

Spurious radiation from an amateur station shall be reduced or eliminated in accordance with good engineering practice. This spurious radiation shall not be of sufficient intensity to cause interference in receiving equipment of good engineering design including adequate selectivity characteristics, which is tuned to a frequency or frequencies outside the frequency band of emission normally required for the type of emission being employed by the amateur station. In the case of A3 emission, the amateur transmitter shall not be modulated to the extent that interfering spurious radiation occurs, and in no case shall the emitted carrier wave be amplitude-modulated in excess of 100%. Means shall be employed to insure that the transmitter is not modulated in excess of its modulation capability for proper technical operation. For the purpose of this section a *spurious radiation is any radiation from a transmitter which is outside the frequency band of emission* normal for the type of transmission employed, including any component whose frequency is an integral multiple or submultiple of the carrier frequency (harmonics and subharmonics), spurious modulation products, key clicks, and other transient effects, and parasitic oscillations.

Key Clicks

Whenever the key is operated, it either makes or breaks current flow. This, in turn, means that small sparks occur. They may be so minute as to be invisible, but they're there and you can hear them on another receiver, just as you can hear a "pop" in a receiver when you turn on or off a light fixture. A similar situation, except that it's much worse, is called "key clicks"; this occurs when the signal comes on or off too abruptly, and makes the same kind of pop. The difference is that key clicks caused by "too hard" keying are radiated over the same wide range as the signal itself/and therefore are frowned upon by the FCC monitors.

Carrier Frequency & Continuous Wave

Carrier frequency is defined as the frequency at which a transmitted signal is received. When an operator transmits a continuous signal by holding his telegraph key down the emitted signal is referred to as a carrier. That carrier may be modulated, as described in Chapter 5, by a variety of means. As far as the Novice licensee is concerned, the only modulation is achieved by transmitting signals composed of short pulses of *continuous waves*.

D.2. Classification of Emissions

The FCC learned a long time ago that it takes a lot of extra time and effort to spell out the modes of emission each time one of them is referenced. To facilitate describing the modes, the government has developed a special alphanumeric code. They are listed in the table of Fig. 7-10.

EMISSION CODE	DESCRIPTION	TYPE OF MODULATION
AØ FØ	On the air carrier of an rf transmitter with no modulation	
A1	Telegraphy without the use of a modulating audio frequency, (by on off keying) Referred to as "carrier-keyed telegraphy."	
A2	Telegraphy by on-off keying of an amplitude modulating audio frequency or audio frequencies, or by the on-off keying of the modulated emission (special case: an unkeyed emission amplitude modulated).	Amplitude
A3	Radiotelephony	Amplitude
A4	Facsimile	Amplitude
A5	Television	Amplitude
F1	Telegraphy by frequency-shift keying without the use of a modulating audio frequency.	Frequency
F2	Telegraphy by the on-off keying of a frequency modulating audio frequency or by the on-off keying of a frequency-modulated emission (special case: an unkeyed emission frequency modulated)	Frequency
F3	Radiotelephony	Frequency
F4	Facsimile	Frequency
F5	Television	Frequency
P	Pulse	
Special Note: In the amateur radio service, A3 also includes single- and double-sideband transmission with full, reduced, or suppressed carrier.		

Fig. 7-10. Alphanumeric emission mode identification.

D.3. Sample Question

The symbol A1 designates

A. *the purity of an emission.*
B. *the readability of a signal.*
C. *the power level of an emission.*
D. *the stability of an emission.*
E. *the type of emission from an amateur station.*

Electromagnetic Compatibility

This section is on a very sensitive subject, and how well you learn it will possibly save your relationships with the neighbors. Let's face it (we have to whether or not we *want* to!), amateur radio transmitters *sometimes* produce interference with consumer electronics products (and other devices). Note, however, my deliberate avoidance of the word "causes" in the preceding sentence. Amateur transmitters sometimes *do* cause interference to a TV receiver, Hi-Fi set, etc. But it is not always the amateur's fault; it is sometimes due to defects in the equipment being interfered with!

TVI is sometimes very easy to cure. Other times, it is a whole lot like trying to nail Jello to the wall . . . you never seem to get it done right. The only thing that you can do is to make sure that your transmitter is above reproach! There are certain standards of good technical practice for transmitters that will eliminate the possibility that your transmitter is producing the problem.

The principle component of any form of TVI problem is harmonics from the transmitter. The harmonic, you may recall, is an integer multiple of the transmitter frequency. If, for example, you transmitting on 21.1 MHz, you will have a second harmonic at 2 × 21.1, or 42.2 MHz. Similarly, the third harmonic will be at 3 × 21.1 , or 63.3 MHz. Both the second and third harmonics of the Novice 15-meter band will tear up television channels (but for slightly different reasons . . . the second harmonic is a problem due to certain design choices made by TV receiver engineers). If you are on a lower band, then the second harmonics (etc) may not interfere with TV, but will interfere with someone. If the FCC spots your harmonics (and they *do* know where to look!) , they will issue a Notice of Violation.

There are several steps that you can take to make the harmonics very, very weak . . . and therefore less troublesome. First, make sure that you are operating the transmitter with all of the shielding in place. Commercially made amateur transmitters usually have pretty decent RF shielding, and it is part of the cabinet design. Make sure that none of the shielding members are loose or missing. On older equipment, corrosion can make a difference in the effectiveness of the shielding. A little cleaning will do wonders in this case.

Also make sure that the transmitter is well grounded. This means at least one (preferably more!) *8 foot* ground rods driven into the ground close to your transmitter location (if the transmitter is on the second floor, then you might have a problem). The wire, preferably braid or strapping material, between the ground rod and the transmitter should be wide (to reduce its inductance) and as short as possible. The shortness means less than quarter wavelength. This is not too much of a problem on 80-meters, or even 40-meters, but in the 15- and 10-meter amateur bands the length "gets long in a hurry." In my own shack (K4IPV), I have a seven-inch wide copper strap between the operating table and the ground system. That ground system consists of six 8-foot ground rods (copper clad steel) connected to the copper strapping. The total run between the table (at its farthest end) and the ground rods is approximately six feet.

A TVI filter is also called for. These filters are *low pass filters*, and have a cut off frequency above 30 MHz. They will pass the amateur signal, but will not pass any harmonics above 30 MHz (attenuation is 10,000:1 at 40 MHz in one popular model. This means that the second harmonic of a 15-meter novice station is more than 80 dB down from the main carrier). TVI filters cost about $20 for a 250 watt model and $30 for a 1000 watt model . . . and are well worth the money.

An antenna tuner is another method for reducing the harmonic output of the transmitter. These devices are normally used to provide a match between the transmission line, or transmitter output, and the antenna. They are also useful for reducing the harmonics. This latter application comes from the fact that the tuner is one more tuned network that will pass the carrier frequency, while attenuating all other frequencies (including those pesky harmonics!). Many manufacturers now produce coax-to-coax antenna tuners that are ideal for most amateur stations (they will "match" an impedance ratio of 1:1, providing some additional frequency discrimination). The tuner should be placed at the output of the low pass filter, which in turn, is placed at the output of the transmitter. The placement of equipment in an amateur station will be given more complete consideration in a later section of this chapter.

When you have a well-shielded, properly grounded transmitter that uses a low-pass TVI filter and an antenna tuner between the transmitter and the antenna, you are well covered and should not have any problems that are your fault. There is, however, one additional thing that you can do: use resonant antennas. Some antennas can be used on a wide range of frequencies, especially if an antenna tuner is available to match the impedance of the antenna to that of the transmitter output. But simple resonant antennas (dipoles, verticals, etc.) act like an LC tuned resonant circuit! This forms one last barrier to those harmonics! The dipole or resonant vertical will not pass most of the harmonics, especially those that are even numbered (some antenna *will* pass the third and fifth harmonics disgustingly well!).

The TVI is your fault only when your transmitter is producing harmonic energy (which, incidentally, is illegal) sufficient to cause interference with nearby receiver. You are on shaky legal grounds if your transmitter lacks any of the anti-harmonic devices mentioned above. But, should you have all of these things, and interference still exists, your neighbor simply has to get his set fixed.

A principle cause of TVI, that is not the fault of the transmitter, is overload of the TV or radio receiver. In that case, the TV set will create harmonics of your signal where none existed before! These harmonics are then treated like any other in-band signal by the TV . . . and TVI is the result. The best way to get around this problem is to install a high-pass filter at the antenna terminals of the receive (preferably on the tuner inside of the TV). The high pass filter will pass the TV signal, but attenuates the amateur signal. BUT, YOU SHOULD NOT INSTALL THE FILTER. Once you touch the neighbors TV, you are married to it! Any time something goes wrong with that set from now will be *your fault*. Your neighbor might be boorish enough to suggest that you pay for any and all future service calls (after all, it was *your* transmitter that . . .). In still other cases, the neighbor hasn't taken an assertiveness training course (like combining public speaking and Karate lessons in the same semester), so will merely mutter behind your back . . . to all of the other neighbors. Have the neighbor hire a professional TV repairman to install the filter.

The FCC has published a booklet on TVI, and it can be bought for a small fee: *How to Identify and Cure Radio-TV Interference*.

The booklet can be purchased from the Consumer Document Center, Pueblo, Colorado, for $1.50. The advantage of this book is that it points out gently that the problem could be the fault of the complainant's TV or radio, and not the amateur's transmitter! After all, people do not ordinarily realize that the TV has two functions: it must respond properly to valid TV signals, *and* it must not respond to signals on other frequencies . . . that includes your amateur transmissions!

E. ELECTRICAL PRINCIPLES

E.1. Definitions

(1) *Electromotive force*
(2) *Resistance*
(3) *Capacitance*
(4) *Inductance*
(5) *Alternating current*
(6) *Kilohertz (kHz)*
(7) *Direct current*
(8) *Voltage drop*
(9) *Electrical power*
(10) *Coulomb*
(11) *Rectification*
(12) *Megahertz (MHz)*

The definitions for these terms are included in Chapter 1 *Learning the Language*.

E.2. Fundamental Units and Concepts

(1) Units
- *(a) volt*
- *(b) ampere*
- *(c) ohm*
- *(d) joule*
- *(e) watt*
- *(f) henry*
- *(g) farad*

(2) Potential difference
(3) Electric current

Fundamental Units

These are covered in Chapter 1, but we'll bring them all together here to make it easy for you to see them all in perspective.

The *volt is the unit of electromotive force*, or potential difference. The *ampere is the unit of electric current*, and the *ohm is the unit of electrical resistance*. It takes one volt of potential difference to cause a current flow of one ampere through a resistance of one ohm. The *product of voltage and current is power, the unit of which is the watt*; so when one ampere of resistance flows through one ohm, it not only means that one volt is the electromotive force applied but that a power of one watt is being used in the process. The *joule is the unit of electrical energy, the henry is the unit of inductance, and the farad is the unit of capacitance*.

E.3. Direct Current Theory

(1) Ohms law

(a) Resistance in series

(b) Resistance in parallel

(c) Series-parallel circuits

(2) Characteristics of parallel circuits

(d) Voltage division

(e) Power

(f) Conductance

(2) Characteristics of parallel circuits

(a) Current through the branches

(b) Voltage across the branches

DC Theory

There are three basic types of circuits:

- Series circuits.
- Parallel circuits.
- Series-parallel circuits.

In a series circuit the same current passes through each device in completing its path to the source of supply.

In a parallel circuit the same current does not flow through each device in completing its path to the source of supply; the current divides to follow two or more parallel paths.

A series-parallel circuit is a combination of series and parallel circuits.

There is a rise of potential in the direction of (electron) current flow; a fall of potential in the opposite direction.

The voltage drop between two points is the potential difference required to produce the current between the two points.

Voltage drops across resistances are called IR drops, since they are computed from the Ohm's law formula, $E = IR$.

In a simple series circuit, the sum of the voltage drops in the external circuit is equal to the applied voltage.

In a parallel circuit the same voltage is applied to each parallel branch.

In a parallel circuit the total current is equal to the sum of the currents in the individual branches.

In a parallel circuit the effective resistance of the parallel branches is equal to the voltage applied to the branches divided by the total current through the branches.

In a series circuit, the effective resistance is equal to the voltage divided by the current.

Parallel resistance may be combined to obtain the effective resistance by adding the reciprocals of all resistances and converting the sum to a reciprocal.

Any circuit may be reduced to an equivalent series circuit.

The sum of the currents arriving at any point in a circuit is equal to the sum of the currents leaving that point.

E.4. Principles of Magnetism

(1) Fundamental laws

(2) Magnetomotive force

Faraday, in his now classic experiment, connected a sensitive galvanometer across a coil and found that when a magnet was thrust into the coil a current flowed in the coil, and that when the magnet was withdrawn a current flowed in the opposite direction. Current flow, however, resulted only during the time the magnet was *moving*—that is, when the lines of force about the magnet cut the wires of the coil. The opposite condition was also found to be true, that if the magnet was held stationary and the coil moved, current flowed during the time of moving. Thus, an alternating current had been produced. Figure 7-11 illustrates this principle in terms of a single conductor and a horseshoe magnet.

Inserting a soft iron core into a solenoid greatly increases the number of magnetic lines of force. This increase in magnetic lines is not from an increase in the intensity of the field, which depends only on current and turns per unit length, but from additional lines produced by the magnetization of the iron core.

Because magnetic flux or lines of force form closed loops, the path that the flux loops follow is called the magnetic circuit. Electrical circuits and magnetic circuits have many points of similarity. The force which produces a flow of electrons in an electrical circuit is the electromotive force. In the magnetic circuit

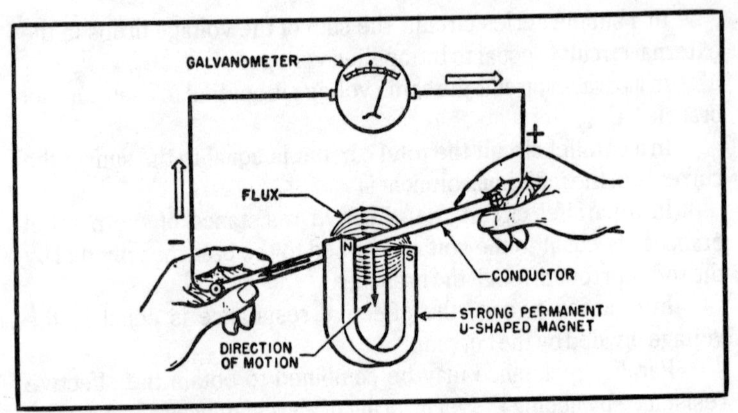

Fig. 7-11. Inducing an emf.

the force which produces the flux is called the magnetomotive force. Similarly, just as the resistance opposes the flow of current in an electrical circuit, so *reluctance* opposes the magnetic flux in a magnetic circuit. Similarly, just as *conductance* indicates the ease with which electrical current flows, so *permeability* indicates the ease with which magnetic lines of force flow in a magnetic circuit.

No name has been given to the unit of reluctance in a magnet circuit. Quantitatively, one unit of reluctance is the reluctance of a magnetic circuit one centimeter long and one square centimeter in cross section with permeability. Mathematically, the unit of reluctance is expressed by the equation,

$$R = \frac{1}{\mu A}$$

where 1 represents length in centimeters, A the cross-sectional area in square centimeters, and μ the permeability.

In much the same manner that Ohm's law expresses the relation between current, voltage and resistance, the expression,

$$\phi = \frac{\text{mmf}}{R}$$

gives the relation between magnetomotive force, flux and reluctance where flux is in maxwells, *mmf* the magnetomotive force in *gilberts* (a gilbert is the mmf required to establish a flux of one maxwell in a magnetic circuit in which the reluctance is one unit), and R the reluctance.

The magnetizing force set up due to current flowing in a coil or solenoid is known as the magnetomotive force. Since the strength of

the magnetic field about a conductor increases when the current through the conductor increases, the magnetic field about a coil or solenoid will also increase when the current through the coil is increased. In fact, for any coil, if the current is doubled the strength of the field will also double. Also, since the total magnetic field above a coil is a summation of the field of the individual loops or coils, if the number of loops is increased, the strength of the magnetic field will increase. Therefore, the amount of flux (lines of force) about a helix, whether it has an air core or an iron core is proportional to two factors:
(1) The current in amperes flowing in the coil.
(2) The number of loops or *turns* in the coil.

The word *turn* means just one wrap of a conductor around a core which may be either air for solenoid or in the case of an electromagnet, a piece of soft iron (A of Fig. 7-12).

The magnetomotive force is proportional to the current (in amperes) in the circuit and to the number of turns of the coil.

E.5. Sample Question

The electromotive force (emf) that will produce a current of one ampere through a resistance of one ohm is a

A. henry
B. farad
C. joule
D. volt
E. watt

F. PRACTICAL CIRCUITS

F. I Basic Circuits

(1) Battery with internal resistance and resistive load
(2) Elementary oscillator circuit
(3) Elementary amplifier circuit

Internal resistance

If a voltmeter is connected between the electrodes of the voltaic cell shown in A of Fig. 7-13, a reading of 1.08 volts, the open-circuit voltage of the cell, will be obtained. That is, 1.08 volts is the emf produced between the two electrodes when the cell is not delivering current to an external circuit.

If a resistor is also connected across the terminals of the electrodes, for example, the 2.8-ohm resistor in B, current will

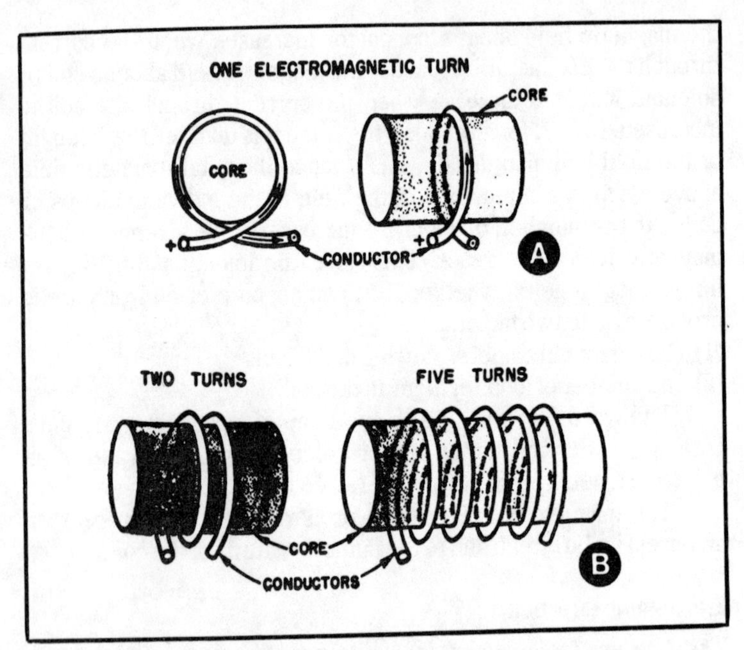

Fig. 7-12. Electromagnetic turns.

flow from the negative to the positive electrode, as indicated by the arrow. The current which leaves the cell at the negative electrode and the current which enters the cell at the positive electrode must flow through these electrodes, which have a small amount of resistance. Furthermore, zinc ions have to be pushed into the solution at the zinc electrode, and hydrogen ions have to be pushed out of the solution at the copper electrode. The electrolytic solution presents resistance to the motion of the ions. Therefore, it takes force to make these ions move. A part of the total emf of the cell is expended in making the ions move inside the cell.

This behavior of the cell is precisely that of resistance, so that the cell is said to have *internal resistance*. Suppose that the internal resistance of the cell in B is 0.2 ohm. The cell can then be represented as a theoretical 1.08-volt cell which has no internal resistance in series whith a resistance of 0.2 ohm (C of Fig. 7-13). Since the external resistance in the circuit is 2.8 ohms, the total resistance in the series circuit is 0.2 plus 2.8, or 3 ohms. The total current in the circuit is

$$3 = \frac{E}{R} = 10.8$$

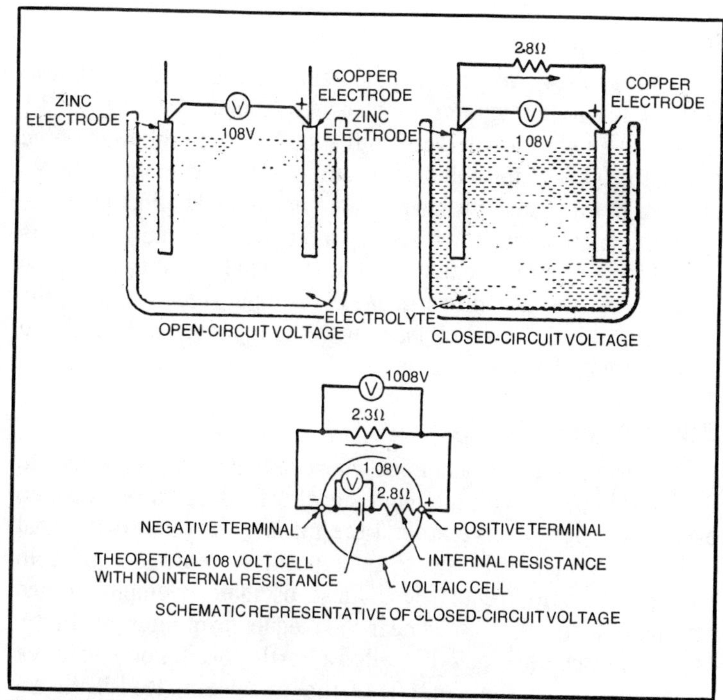

Fig. 7-13. Effect of internal resistance on cell voltage.

0.36 ampere, and the voltage drop across the internal resistance of the cell is—

$$IR = 0.36 \times 0.2 = 0.072$$

The voltage or potential difference across the 2.8-ohm resistor is—

$$IR = 0.36 \times 2.8 = 1.008 \text{ volts}$$

This is the voltage at the terminals of the voltaic cell.

By definition, the *open-circuit voltage* of a cell is the voltage between the terminals when no appreciable current is flowing. This voltage is measured by placing a voltmeter across the terminals of the cell when it is not delivering current. The voltmeter, because of its high resistance, draws so small a current to deflect its needle that there is said to be practically an open circuit between the terminals.

The *closed-circuit voltage* of a cell is the voltage between its terminals when current is flowing through an external circuit (B of Fig. 7-13). The *terminal voltage* of a cell is merely the voltage between its terminals (electrodes) and can therefore be either the open-circuit or closed-circuit voltage.

Elementary Amplifier

In Fig. 7-14 two very common amplifier circuits are shown. On the left is a common-emitter transistor stage; on the right is the vacuum-tube equivalent. The input signal is fed to the base of the transistor (grid of the tube). The output circuit contains a series element in the transistor made up of emitter and collector; in the tube, the output circuit is a series path of cathode and plate. It is important to remember that the output signal, developed across the *load* resistor, is a mirror image of the input signal—that is, the ouput signal is 180° out of phase with the input signal in both the sketched amplifiers.

Elementary Oscillator

An oscillator is an amplifier whose output signal is fed back to the input in phase with the input signal. The differences between oscillator types revolve around the manner in which that signal feedback is obtained. Since most oscillators are designed to operate at radio frequencies, most oscillators employ tuned circuits that resonate on or near the frequency of interest. In the oscillator circuit of Fig. 7-15 (called a Hartley oscillator), feedback is obtained from the emitter of the transistor (lead with arrowhead). If capacitor C_E were removed, the circuit would not oscillate at all. This capacitor acts as a short circuit to alternating current, thus effectively bypassing rf signal components around the emitter resistor R_E.

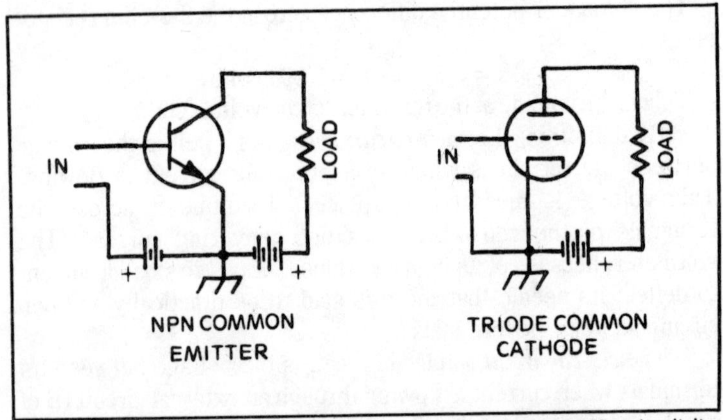

Fig. 7-14. The common emitter is the most popular transistor amplifier circuit. Its tube-type equivalent is the conventional common-cathode arrangement. The common-emitter circuit has the highest gain of all the amplifier types. The output waveform is an inverted duplicate of the input signal.

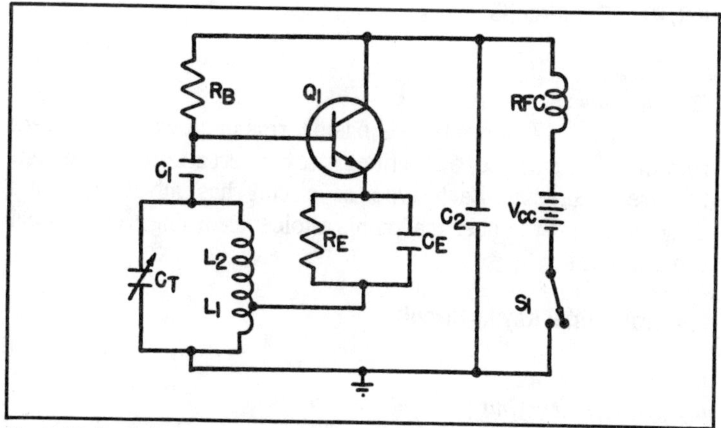

Fig. 7-15. Transistor Hartley oscillator.

F.2. Filter Circuits

(1) Lowpass

(2) Highpass

A high-pass filter does what the name implies: it allows frequencies higher than the filter's design frequency to pass, but blocks signals whose frequency is below the design frequency. A low-pass filter passes signals of lower frequency than the filter's design frequency. Figure 7-16 illustrates both high-pass and low-pass filters.

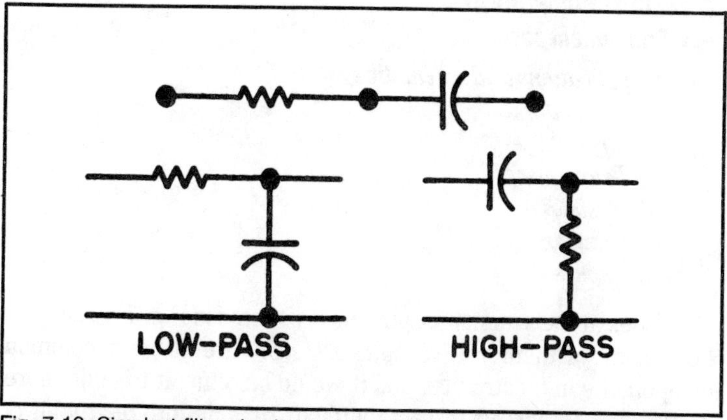

Fig. 7-16. Simplest filter circuit consists of one resistor and one capacitor in series, as shown at top. This forms an RC L-section which may be used for either low-pass or high-pass purposes as shown in the two lower schematics. Method by which filter operates is explained in text; all filters work on this basic principle, but more complicated circuits modify it in various ways because they contain not just one but several reactances, each of which behaves differently.

F.3. Rectifier Circuits

(1) Half wave
(2) Full wave

Figure 7-17 shows the six basic forms of alternating-current rectifier. The purpose of a rectifier is to convert alternating current to direct current. Each type of circuit has advantages and disadvantages; the user makes his choice according to individual requirements.

F.4. Transmitter Keying Circuits

(1) Cathode keying
(2) Grid circuit keying
(3) Screen grid keying

Keying circuits are described in Chapter 6.

F.5. Sample Question

What is the function of the circuit shown in Fig. 7-18?

A. Two-voltage power supply
B. Full-wave rectifier power supply.
C. Voltage regulator.
D. Full-wave voltage doubler.
E. Choke-input filter.

G. CIRCUIT COMPONENTS

G. I. Component parts

Types, applications and schematic symbols

(a) Capacitors
(b) Crystals
(c) Transformers
(d) Resistors
(e) Inductors
(f) Chokes

Look at the schematic symbols presented in Fig. 7-19 and see how many of them you recognize. These are all very common components in electronics, and it would be difficult to build more than but a few circuit projects without running into nearly all of them. If you've been building projects in the past, you have already had the chance to work with the resistor (A), the potentiometer B, the capacitor D and E, the electrolytic capacitor I, the switch P, battery R, the momentary-contact switch U, the chassis ground

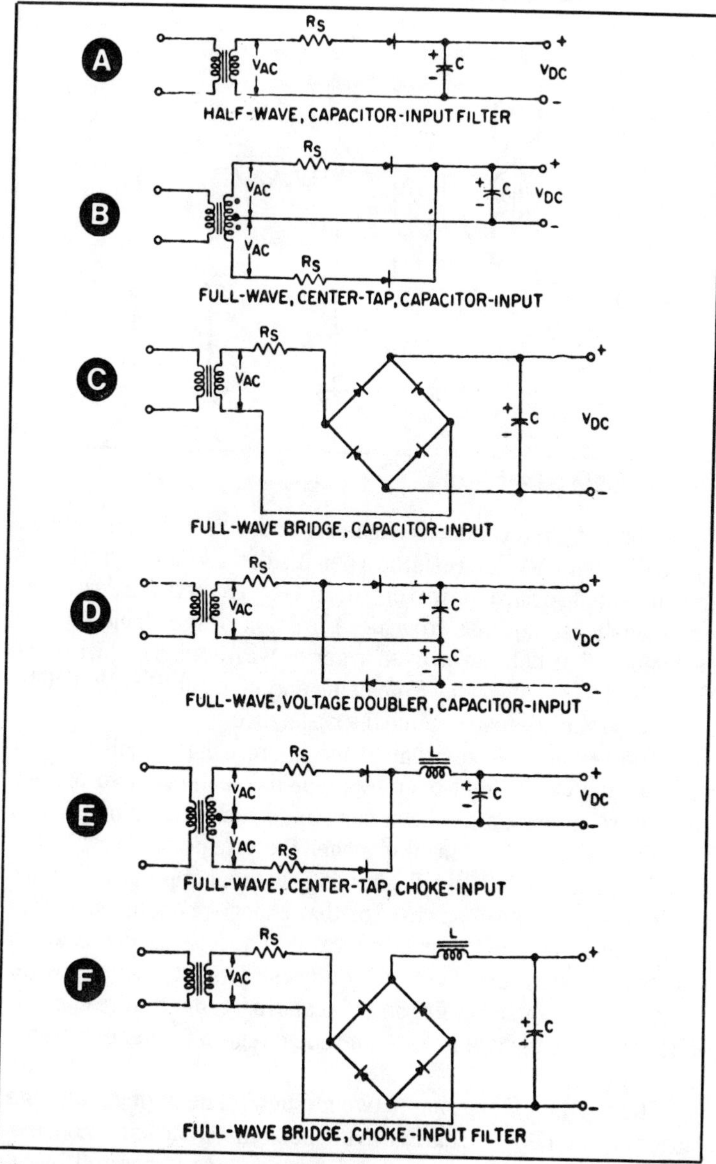

Fig. 7-17. Basic semiconductor rectifier circuits.

(X), and the tie point (Y). You may recognize some of the others, but there are probably a few that you haven't seen before.

C, for example, is a photoresistive cell. The device looks very much like the symbol—a circular field on which is imprinted a

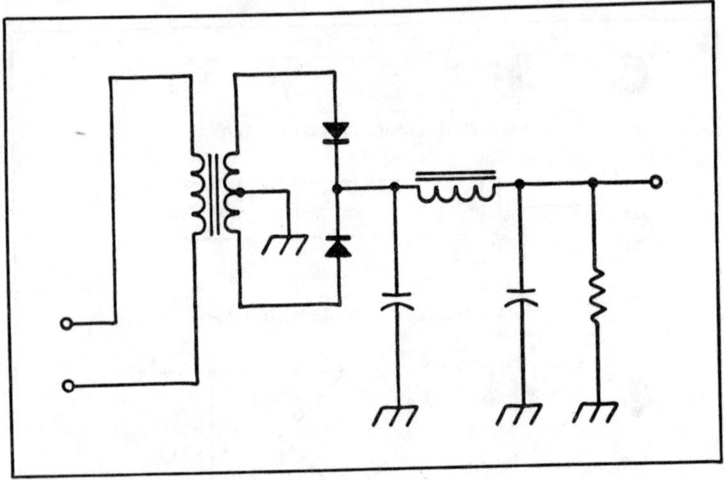

Fig. 7-18. (For Question F.5.)

zigzag line of photoconductive material, such as cadmium sulfide. The cell is a high-value resistor when it's dark. When light strikes the imprinted surface, however, the device's resistance decreases enormously, sometimes to a mere fraction of the device's dark resistance. The cell can be used in series with a relay coil wire, for example, to keep the relay from being energized until enough light strikes the surface to overcome the resistance.

The two arrows adjacent to the photoresistive cell tell you that the device is light-operated; you'll see those two arrows adjacent to transistors, silicon controlled rectifiers, diodes—and they always mean the same thing when they are pointed toward the symbol. The resistive-cell drawing without the two arrows signifies a thermistor—a resistor that changes conductivity with temperature. Some thermistors exhibit a marked decrease in resistance when they get hot; others do just the opposite. Thermistors are used as fire and heat alarms, temperature control, and for voltage control in vacuum-tube circuits, to mention only a few applications.

Items F and G are simply two methods of depicting a variable capacitor. You learned earlier in the book that a capacitor consists of nothing more than a couple of metal plates separated by an insulator, or dielectric. The distance between the two plates, or the dielectric strength of the insulation material, determines the value of the capacitor. A variable capacitor allows the distance between the plates to be varied manually, normally between two specified values.

Item H is a piezoelectric element, normally referred to as a "crystal." Depending on its thickness, size, grain of its axis, a crystal can produce a tiny alternating voltage, and when a pressure is applied, it vibrates physically. The rate of vibration is directly proportional to the variations in applied pressure. The crystal

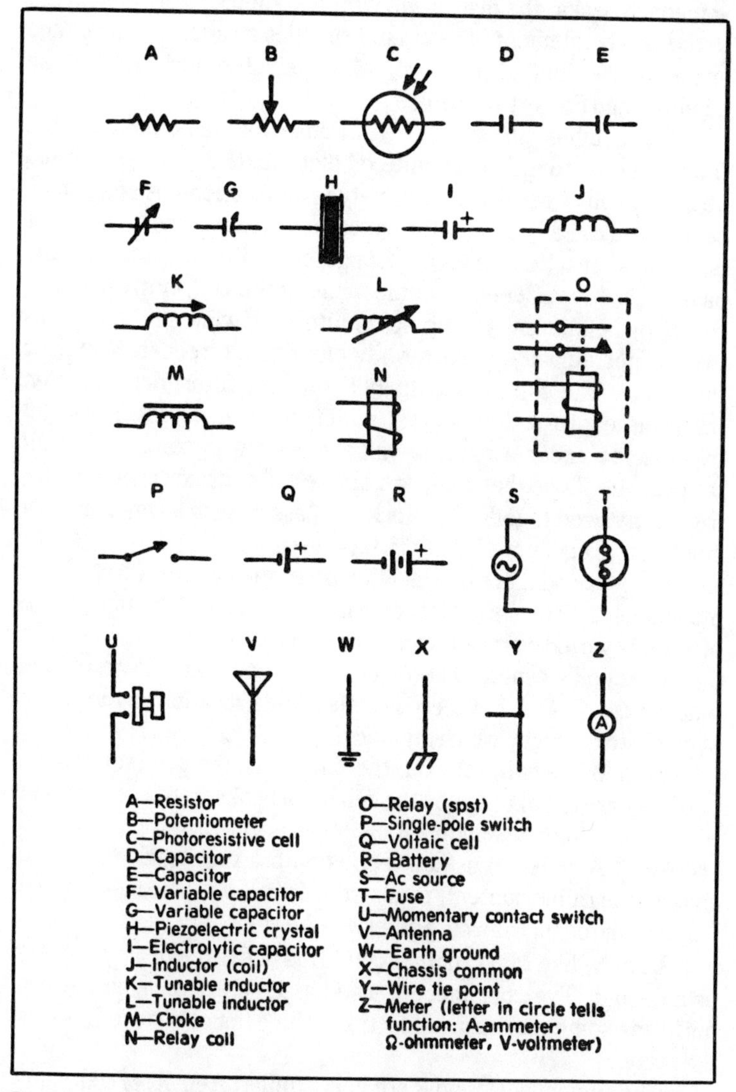

Fig. 7-19. Popular schematics symbols. How many do you recognize. Cover the lower portion of the page and try to name as many as you can. If you can't name more than 15, you'd better study the explanatory chart.

element as shown in the symbol is used in radio circuits because the output is very precise and stable.

Though the symbol for a crystal microphone is different from the one pictured in, the operation is exactly the same: A diaphragm is mechanically attached to the piezoelectric element. When someone speaks, the diaphragm vibrates, thereby applying stress to the crystal element. The crystal supplies an alternation voltage analog of the applied pressure, which is subsequently fed to an amplifier and finally to a speaker.

The symbols shown in J, K, L, and M are all similar in that they are coils. Item J is a simple coil that could be of any length and be used in any one of a large variety of applications. A coil can be considered roughly to have the opposite function of capacitors; that is, while a capacitor blocks dc and passes ac, the coil blocks ac and passes dc. Much depends on the value of the coil, however. The capacitor is inherently a high-pass filter, allowing all frequencies over a specific value to pass while blocking all frequencies under that value. The coil is essentially a low-pass filter; depending on the value expressed in henrys, millihenrys, or microhenrys , all frequencies under a specific value are allowed to pass, but all frequencies above that value are blocked. Coils and capacitors are frequently used together to block and pass in such a way that very stable one-frequency signals can be generated.

Items K and L are one and the same; they define a coil whose inductance may be varied by some means. Usually, the inductance of a coil is varied by means of a ferrite core that is screwed into and out of the coil's center. There are several methods of varying the inductance of a coil, but there is little standardization on the use of symbols to depict by which method the coil referenced is varied.

Item M is a coil, too. But the line above the coil (sometimes two lines are used) tells you the coil has a husky iron core. Coils of this type often resemble transformers, and they are called "chokes." A choke is used in series with a varying dc source to smooth the ripple content, turning the dc into a purer, more stable voltage source than it would otherwise be.

Item N, like M, is another coil. This one is the heart of a relay or solenoid. The core piece is used as an electromagnet, which performs some mechanical function. The symbol next to it (O) is the relay.

The symbols of Q and R are very similar. Item R, of course, is a battery. But a battery consists of more than one individual cell. The cell itself is represented by the symbol shown in Q. Notice the marked resemblance between the single cell and the capacitor, as depicted in schematic diagrams. The capacitor consists of two parallel lines of equal length, however, while the cell consists of parallel lines of different lengths. The longer of the two parallel lines is always the positive terminal of the cell. Similarly, the battery symbol contains parallel lines of two lengths. The end with the longer line is always positive. If the line on each end of the battery is of the same length, the symbol has been incorrectly drawn.

There are some similarities between the symbols of S and T, but the similarity stops with the artwork. The horizontal S in the circle always represents an ac voltage source; it may mean a household outlet or it might mean an ac generator. Sometimes it is used to depict a sine-wave signal generator. The only thing you know for sure when you see this symbol is that an ac voltage is generated in the place where the circle is used. The other symbol, an S that is connected to the schematic diagram's power lines, depicts a fuse. It could be used in ac or dc circuits, and might be any value, from a few milliamperes to thousands of amperes; usually the value is printed adjacent to the symbol.

The triangle shown in V is an antenna. The fact that it is shown as a triangle does not mean that a specific configuration of antenna must be used. As a matter of fact, more often then not an ordinary piece of wire very long, of course will suffice, particularly in receiver circuits. If there is some particular requirement with regard to the type of antenna, its construction, etc., there will be a note on the diagram so stipulating.

The symbols of W and X have caused more furor in the electronics industry than any of the others—perhaps even more than all of the others put together. There is nothing complicated in them, I hasten to add; it's just that W was at one time used to denote what X depicts today. And instead of eliminating the symbol of W entirely, it was reassigned to depict another related function. The three horizontal parallel lines of diminishing size refer today to an earth ground. If you see this symbol on a modern schematic diagram, you will know that means you are to connect the wire to a waterpipe, conduit, or other conductor that is physically sunk into the earth outside your home. In other words, it means "ground" in the most literal sense of the word.

The other symbol X is used to mean "ground" also; but when this symbol is shown, you need not make certain there is an actual earth-ground connection. You can use one heavy bus wire in whatever circuit you're building, and tie all "grounds" to this common line. If you're working with metal chassis, the chassis itself will undoubtedly be used as the common tie point. It is always good engineering practice, of course, to electrically ground all chassis to the earth anyway, so the differences between the two symbols aren't monumental; nonetheless, the "rake" ground, as it is called, simply means: "tie all points marked this way to one common bus wire."

The final symbol in Fig.7-19 is a meter. The letter in the circle will tell you the kind of meter you should use, and it might be anything from a simple voltmeter or ammeter to a complicated instrument such as an impedance-measuring meter.

There are other symbols for passive components, but these and the transformer, of course are the most common. As you work more and more in electronics, you'll get more and more familiar with the symbols in current use. If you learn the ones shown in this subsection and the active ones shown in the next, you're not likely to be "snowed" by any circuit.

G.2. Semiconductor Devices

1 Diodes
- *(a) Types*
- *(b) Applications*

2 Transistors

Semiconductor Devices

It would be possible to include a large section on vacuum tubes, but these devices are being replaced more and more with transistors of one type or another, and the field of vacuum-tube devices gets more restrictive by the day. Instead, we concentrate on the more popular of the many available semiconductors. In Fig. 7-20, 22 schematic symbols are shown; some of these symbols are redundant—the tunnel diode, for example, may be depicted by either one of two schematic drawings. Look at the symbols before reading further; see how many you can recognize without looking at the explanatory text that accompanies the drawings. Try to familiarize yourself with the basic diode sketches, and see how an additional mark or line can change the meaning of the symbol.

Bipolar Transistors

Sketches A,M, N, T, and U are all representative of bipolar transistors, but what are the differences between them? Items A

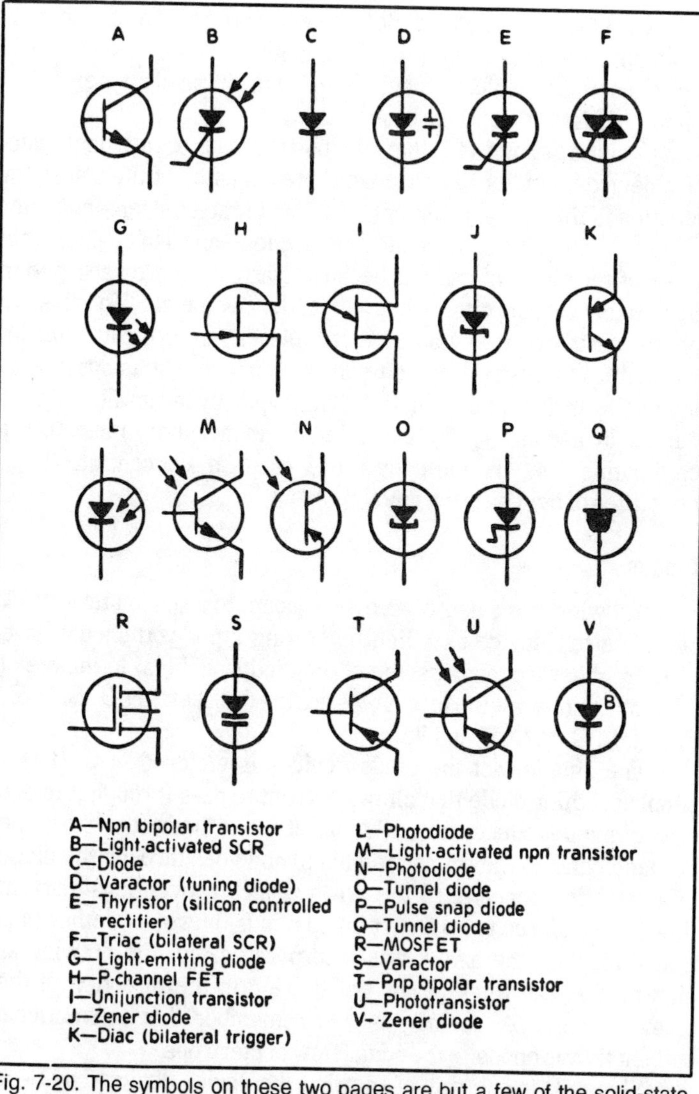

Fig. 7-20. The symbols on these two pages are but a few of the solid-state discrete components in current use in the electronics industry. How many can you recognize? Cover the answers and try to name them. Devices D and S are the same, like two ways of spelling a word; similarly, devices J and V are identical, as are L and N and O and Q.

and T should be relatively easy; A is an npn transistor, and T is a pnp equivalent. The arrow direction on the emitter tells you whether the bipolar transistor is pnp or npn. In npn devices, the arrow points away from the transistor junction; in pnp devices, the arrow contacts the junction. Both M and U are phototransistors; M is an npn, and U is a pnp type. The symbol shown in sketch N is often used to depict a photodiode, but it really refers to a phototransistor that has no base lead.

An interesting fact is that all bipolar transistors are photoelectric devices! One of the peculiar characteristics of the transistor junction is the fact that current will flow in the emitter—collector circuit when light strikes it. The phenomenon isn't observable most of the time, of course, because transistors are packaged in light-tight cases, either black plastic or metal. To make a phototransistor, the manufacturer places an ordinary bipolar transistor in a case designed to allow as much light as possible to fall on the surface of the junction. This typically means housing the device in a clean epoxy case, but in many phototransistors a collimating lens is built into the case to concentrate large quantities of light onto the junction.

Diodes

Knowing that a pair of arrows adjacent to a schematic symbol identifies the device as a light-operating type, you should have little trouble recognizing some of the diode symbols. Members of the diode family are those symbols pictured in sketches B, C, D, E, F, G, J, L, O, P, Q, S, and V.

The simplest of the diode devices is pictured in C; it is a simple rectifier diode that allows current to pass through it in one direction only. Electron flow begins at the cathode (the short line perpendicular to the connecting lines) and goes through the diode in a direction opposite to the arrow. Since conventional current flow is considered to go from plus to minus, however, rather than minus to plus, the arrow was incorporated into the drawing as shown. You can think of current flowing in the direction of the arrow if you like, as long as you remember that conventional current flow is opposite the actual flow of electrons.

Sketch D is one way of showing the device depicted by S—a varactor, or tuning diode. In this device, the junction between the positive and negative impurities is separated by an undoped region of the intrinsic semiconductor material, which serves as a dielectric, or nonconductor. In many ways, the diode acts as a conventional capacitor. However, because of the "negative-resistance" characteristic of the junction under certain applied-voltage conditions, the value of capacitance can be changed in direct proportion to the applied bias voltage. The varactor has many uses because of the negative-resistance characteristic, but the most common is simply as a substitute for a manually variable capacitor.

Sketches B and E are very similar, with one exception: B is light-operated while E is not. Both are thyristors, often called "silicon controlled rectifiers" or four-layer diodes. The basic thyristor, shown in sketch E, is a thyristor of npnp construction. The three terminals are the anode, the cathode, and the gate. The cathode is the short line perpendicular to the connecting line, as it is in the ordinary diode. The anode is shown at the top. The gate element, which triggers the device into conduction, is the line from the side that connects at the symbol's junction. The thyristor is used in dc switching circuits to serve in much the same manner as a relay. Thyristors can handle fairly large amounts of power from cathode to anode, so the device is inserted in a high-current line like a mechanical relay. Until current is allowed to flow, the thyristor remains an open switch. As soon as a small gate current is introduced, however, the anode-cathode "switch" closes, and current flows in the circuit being controlled.

Sketch B operates the same way, except that gate current flows automatically as soon as light strikes the surface of the device's lens. An interesting characteristic of thyristors of both types is that once the anode-cathode circuit closes, it stays closed—even when the gate current is removed entirely. This phenomenon has both good points and bad. A good point is that the electronic switch can be made to trigger with no more than a quick pulse, which makes it kin to the latching relay. A disadvantage, though, is that some means for shutting down the switch must be incorporated in the circuit design. In burglar alarms and similar applications, the thyristor can be shut off by the mere expedient of manually opening the anode-cathode circuit momentarily.

The strange-looking symbol shown in sketch F is officially referred to as a "bilateral thyristor," or, more popularly, a "triac." Operationally, the triac is the same as a pair of thyristors cross-connected, and with a single gate. In practice, the triac allows switching of dc or ac on a nonlatching basis. You can consider the triac, for all practical purposes, as an ac relay. There is no cathode or anode, since the upper and lower terminals are the same: so these terminals are simply numbered 1 and 2. The gate is referenced to terminal 1, and when gate current is caused to flow, the two main terminals "close," allowing high-current loads to be controlled. Since the triac can be made to trigger with lightning-fast speeds, the device is used in lamp dimmers and motor speed control applications by turning on and off the household power at a specified rate.

The symbol shown in sketch G has the two arrows that mark it as a light-operated device, but notice that the arrows point outward rather than inward toward the device's junction. Outward arrows mean the unit actually generates light. This symbol, then, represents the light-emitting diode. When proper dc bias is applied, the device glows a bright red (or green, depending on the structure of the junction). The unit is used extensively in indicators, readouts, and other such applications because it generates a lot of light and very little heat, making it one of our most efficient light sources. Also, since it is of solid-state construction, it has a virtually unlimited lifetime.

Sketches J and V are the same—different ways of showing the zener diode. (The B in the circle of sketch V refers to "breakdown," because it is the zener's breakdown point that determines its value.) The zener is a natural voltage regulator. Beneath the breakdown voltage point, the zener seems much the same as any other simple diode. When sufficient voltage is applied to reach the device's breakdown point, however, the diode will conduct in a reverse direction. Almost without regard to the amount of applied voltage, the voltage drop across the zener will remain the same as the voltage it saw at breakdown. A 12V zener will maintain a cathode-to-anode potential of 12V when more than 12V is applied, and the voltage across it remains 12V no matter how high the input voltage goes. Since the zener must dissipate the excess voltage (over and above the breakdown point), it must be capable of handling the power excesses; otherwise, the heat of the dissipated power will destroy the semiconductor.

Sketch L is a photodiode; it conducts as long as light strikes the surface of the junction. When the light is removed, the diode ceases to conduct. Sometimes the symbol is used interchangeably with the baseless transistor diagram of sketch N.

Sketchs O and Q are two ways of depicting a tunnel diode, a varactor-like device that has a negative-resistance characteristic and can switch fast enough from one state to another to make it ideal for very high frequency and microwave oscillator applications. The pulse magnitude is so high on the output of these devices that they can actually be used as amplifiers.

The pulse snap diode is shown in sketch P. You won't be using these, probably, because they aren't typically found in hobby circuits. They are used where ultrafast switching is required. Typical applications include pulse generation, waveshaping, and the like.

UNIPOLAR TRANSISTORS

A unipolar transistor is a transistor that has but one junction. Both field-effect and unijunction fit this category.

The junction-type field-effect transistor is shown in sketch H. The important point is for you to note the difference between the junction FET and the insulated-gate FET shown in sketch R.

In the insulated-gate FET, there is no direct connection between the gate and the junction. A thin coating of oxide separates the gate capacitively from the junction, which serves to increase the device's input impedance considerably. The insulation is so thin and fragile that even a touch at the gate of an unprotected device can cause a static-electricity discharge that can destroy the device. If you work with insulated-gate FETs (also called MOSFETs), you must take certain precautionary measures to avoid damaging the FET before you even have applied power to the circuit you've built!

The trigger diode (sketch K), popularly referred to as a "diac," is typically used in conjunction with a triac to control ac loads. The diac is characterized by the need for a specific firing voltage, which must be applied to the unit before triggering can take place.

Now you've made at least a passing acquaintance with the more common members of the growing family of semiconductors. Study the sketches and learn the basics of them.

G.3. Vacuum Tubes

(1) Classification by elements
(2) Basic operating principles

Vacuum-tube theory is included in Chapter 4. If you learn the material in that chapter, you'll have no trouble with test questions.

G.4. Meters

(1) Plate voltage meter
(2) Plate current meter

As shown in the component-symbol sketches, meters are identified in schematics by a circled letter. A V in the circle represents a voltmeter; an A represents an ammeter.

When a reference is made to *plate* voltage or *plate* current, the amplifier type being referred to is a vacuum-tube arrangement. The FCC requires Novices to limit their plate power input (final transmitter stage) to no more than 250 watts. One effective method for doing this is to have a plate current meter and a plate voltage meter to monitor the dc power applied to the final rf amplifier. To determine total dc plate power input, simply multiply the reading of one meter by the reading of the other (after converting them to their basic units). If the plate voltage meter indicates 750V and the plate current meter indicates 333mA (0.333A), the plate power input is precisely 250W. Since meters always have a tolerance of error of 3 to 5%, it would be unwise to operate using the values just described.

There are a number of different types of meter "movement," (i.e. the *works* inside of the meter, analogous to a watch movement). The most popular are variations on the permanent magnet moving coil (PMMC) movement of Fig. 7-21. Two types of meter are based on the PMMC movement: *D'Arsonval* and *Taut-band*. Both types are basically similar to each other, except for the manner in which the moving coil is suspended in the field of the permanent magnet.

The coil consists of several turns of wire on a bobbin (coil form), and is placed in the magnetic field of the permanent magnet. When current flows through a coil it will create a magnetic field around the coil, and the field will be proportional to the strength of the current. This magnetic field interacts with the magnetic field of the permanent magnet, and causes the coil to *deflect* (rotate) an amount that is proportional to the strength of the current. Recall that like fields oppose, while opposite fields attract. This means that the meter coil can deflect in either direction. Some meter movements take advantage of this fact and place the pointer (attached to the coil) in the center of the meter face when zero current is flowing. This arrangement allows us to tell not only the strength of the current, but also its polarity. Most meters, however, place the pointer against a low-end stop at the left-hand side of the meter face when the current is zero. Such meters, therefore, will measure current only when the polarity is correct (we must reverse the meter leads to measure currents of the opposite polarity). The pointer will swing up-scale an amount proportional to the strength of the applied current. This will allow us to calibrate the *distance* that the tip of the pointer moves in units of electrical current.

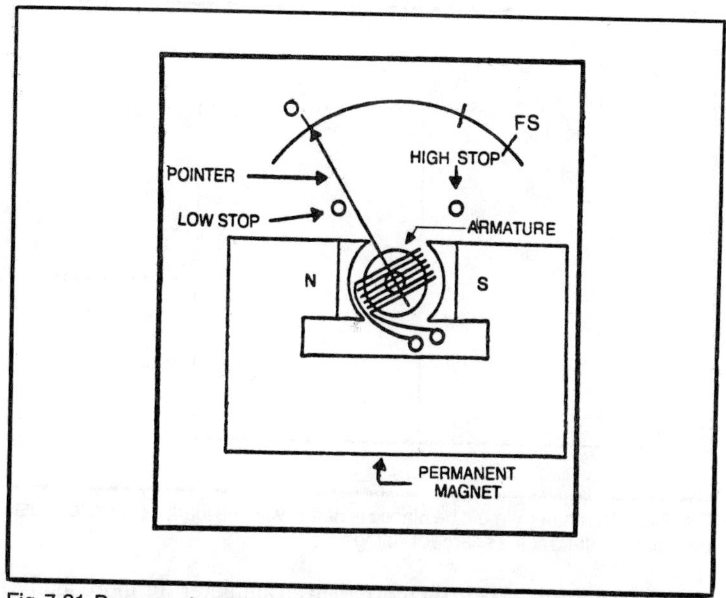

Fig. 7-21. Permanent magnet moving coil (PMMC) meter movement.

The D'Arsonval meter movement is a version of the PMMC scheme in which the bobbin carrying the moving coil is mounted on pivots that are secured to the main frame with jeweled bearings (in the same manner as a watch movement). The taut-band meter movement suspends the bobbin on a metallic ribbon that has been pulled tight (i.e., taut). The D'Arsonval meter movement requires a restoring spring to return the point to zero after the measurement, while the taut-band uses the springiness of the metal band or ribbon for the same purpose.

The basic meter movement is a DC current meter, and is referred to as an ammeter (or milliammeter, or microammeter, depending upon the full-scale current range). A meter to measure electrical current is placed in *series* with the circuit being measured (see Fig. 7-22). The plumbing analogy of electricity, in which the current is likened to the water in a pipe, explains why this is necessary. A voltage is a potential *difference*, which implies that the measurement must be taken *between two points*. This fact tells us that the voltmeter must be placed in *parallel* with the circuit being measured (again, see Fig. 7-22). It is essential that the correct placement of meters be observed, or the meter may be *destroyed*. Placing an ammeter in parallel with a circuit is a sure way to cause its destruction!

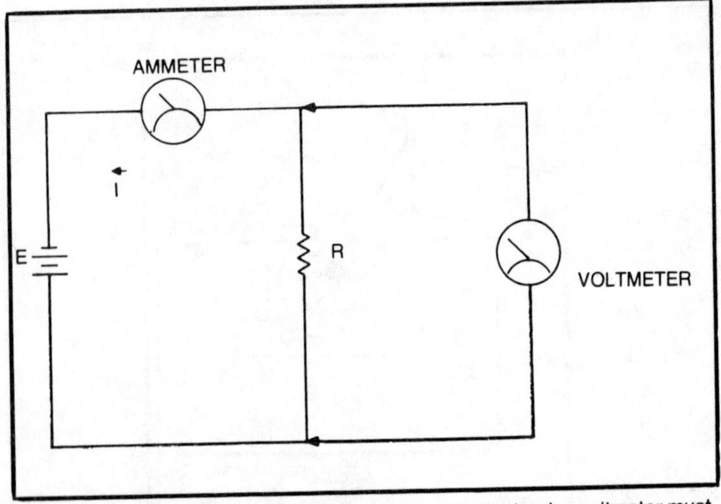

Fig. 7-22. An ammeter must be placed in series with the load; a voltmeter must be across the load or the source of voltage.

The voltmeter is little more than an ammeter (or more likely, a milli- or microammeter) fooled into measuring voltage by reference to current (see Fig. 7-23). We place a ***multiplier resistor*** in series with the current meter, and then allow Ohm's law to take effect. The meter will have some small resistance R_m (remember, the PMMC uses a coil of wire, and wire has resistance), and this resistance must be added to the value of multiplier resistance, R. By Ohm's law, then, we know that the voltage is equal to $(R + R_m) \times I$. The full-scale voltage, therefore, is the product of the full scale current (I_{max}) and the total resistance $(R + R_m)$.

Fuses

Fuses are used to protect amateur radio equipment from overload caused by certain electrical faults. The fuse usually consists of a small section of special wire weakened at one point so that it will melt if the current gets too high. The wire separates when the thin section melts, and interrupts the circuit, thereby shutting off the flow of current.

The fuse protects circuits and equipment (and prevents fires in your home) by turning off the current flow when it reaches dangerously high levels. When a short-circuit occurs in, say, the rectifiers of your transmitter, then it will cause an excessive current to flow in the transformer that feeds the rectifiers. Unless a fuse blows, shutting off the current, you would lose the transformer as well as rectifiers. Similarly, if the transformer develops a short-circuit, then it could set fire to your home *unless* the fuse to the equipment blows *first*. As a back-up, there will be a fuse in your home to protect the house and its wiring from overloads caused by defective equipment. The power company also fuses its lines, so that protection extends to the neighborhood and generating equipment in case of a massive problem in your house (and so on).

There are two lessons you must learn concerning fuses: 1) *Do not defeat the purpose of the fuse*, and, 2) *Fuses don't* **cause** *trouble, they* **indicate** *trouble!* The first lesson applies to all who would use pennies, aluminum foil, and so forth, to short out a blown fuse and make it work again. The second point is the wisdom one author learned from a retired Navy Chief Petty Officer who owned a radio-communications shop. The idea is to find out just why that fuse or circuit breaker blew... don't just replace the fuse and go merrily on your way. There was a cause, and you should find out what it was before trusting the equipment again. Sometimes, in the case of a radio transmitter, the fuse will blow because you keep the plate tank circuit out of resonance too long. The fuse will not pop immediately when the current is higher than the rated value, but there is a time factor. Fuses are generally rated to hold at one current, pop after so-many seconds at a 50% overload, and in a lot less time at a 100% overload. When your plate tank circuit is out of resonance, the fuse is straining, but may not actually blow unless you keep the overload for more than a few seconds. Then the fuse will pop, and you may wonder why.

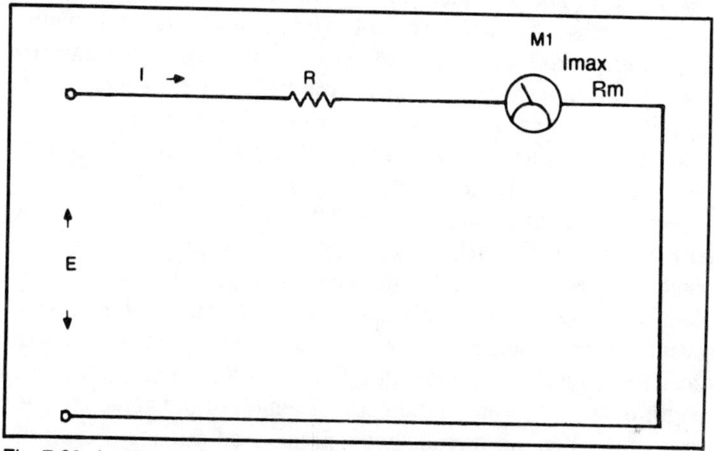

Fig. 7-23. A voltmeter is actually a low-range ammeter with a series resistor to limit the current flow and to provide a useful calibrated scale.

Figure 7-24 shows the circuit symbol for a fuse, and its proper placement in the circuit. It should be the first circuit component from the battery source. Where the fuse is part of equipment powered from the AC-power mains, the fuse is usually placed at a point close to where the power cord from the wall socket enters the equipment cabinet.

G.5. Sample Question

What device is used to rectify alternating current?

A. Choke
B. Capacitor
C. PNP transistor
D. Diode
E. None of the above

H. ANTENNAS AND TRANSMISSION LINES

H.1. Definitions

(1) Electrical length
(2) Field strength
(3) Characteristic impedance
(4) Skin effect
(5) Standing waves

Electrical Length

The length of a given wave depends on the material through which that wave is being propagated, as explained in Section B, *Radio Phenomena*. Despite the fact that the length of a given wave may vary, it is convenient to refer to a frequency by its wavelength. If you were asked to cut a piece of wire to one-quarter wavelength of a given frequency, you'd wonder whether to cut the conductor to the length of a quarter wave in free space or the length of a quarter wave traveling on the wire itself. So the term *electrical length* was inverted. Electrical length is always given in wavelengths.

Radio waves travel at about 66% the speed of light along a length of typical RG-8/U coaxial cable. So a *velocity factor* of 0.66 must be used to calculate the *physical length* of a line whose electrical length is 1 wavelength. A velocity factor of about 5% must be used to calculate the physical length of an antenna whose electrical length is 1 wavelength. For a given frequency, physical length of the wave and the electrical length of the wave are the same only when the wave is being propagated through thin air or empty space. The following table shows the different physical lengths for a signal being transmitted on 7.1 MHz.

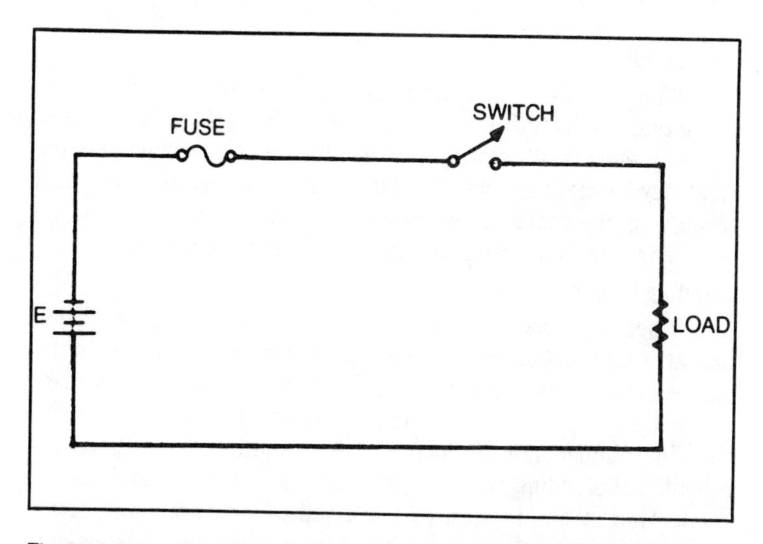

Fig. 7-24. Fuses are to protect the load or the source of power from excessive current flow.

MEDIUM	ELECTRICAL LENGTH	PHYSICAL LENGTH
RG-8/U coaxial cable	1 wavelength	28.17 meters
Long-wire antenna	1 wavelength	40.14 meters
Free space	1 wavelength	42.24 meters

Field Strength

The intensity of a transmitted radio signal is called the *field strength*, and it is measured with a field strength meter. An electromagnetic signal induces a minute voltage on the receiving antenna, which is often less than a millionth of a volt. The higher this voltage as measured somewhere in the radiation field of the transmitting antenna, the higher the field strength is said to be.

Characteristic Impedance

A transmission line has inductance, capacitance, and resistance. Therefore, it possesses the characteristic we call impedance. The characteristic impedance of any line is the impedance value it would have if the line were of infinite length, beginning at a source of alternating current, but extending outward to infinity. In practice, a line is said to be well matched to its load when the resistance at the load is the same value as the characteristic impedance.

Skin Effect

When a conductor carries alternating current, electron motion along that conductor takes place close to the surface. The higher the frequency of alternating current, the more pronounced is the tendency for electron movement to occur at the surface rather than throughout the conductor's entire cross section. This phenomenon is called *skin effect*, and it is not applicable to direct current.

Standing Waves

When the tank circuit of a transmitter feeds a transmission line at a zero-voltage point, and the transmission line feeds the antenna at a zero-voltage point, and the voltage is at its maximum value right at the tips of the radiating elements and the transmission line's impedance is constant over its entire length, there can be but one standing wave on the transmission line—and that is good. This means that every wave fed to the antenna is being radiated, and no waves are reflected back down the transmission line toward the transmitter. This theoretically perfect condition is expressed as a standing-wave ratio of 1:1; that is, for every wave put into the transmission line, a wave is radiated into free space by the antenna.

Such a ratio is impossible to achieve in practice because of the law of diminishing returns. The radiators are not 100% efficient, nor is the transmission line. But while certain amounts of unavoidable losses are to be expected, it is possible to achieve a standing-wave ratio that is close enough to 1:1 that deviations may be ignored.

To understand the mechanism of standing waves, look at the sketch of Fig. 7-25. Since the current at the tip of the antenna must be zero, it must be at its maximum value 90° in from that point, which in this case is the feed point of the antenna, as shown. But consider what would happen if the antenna elements were significantly shorter than a quarter wavelength. The current would drop to zero before the voltage builds up to its maximum value. So the voltage has to build up in the opposite direction—in effect, being reflected back toward the point where the new current and voltage minima and maxima coincide. Since the reflected signal is of lesser magnitude than the input signal, complete cancellation does not take place, so some of the signal does get radiated into space. But the efficiency of the system is degraded considerably.

Any mismatch in impedance anywhere in the signal's path has the same effect, causing reflected signals which interfere with input signals. In practice, antenna tuners are often used at the

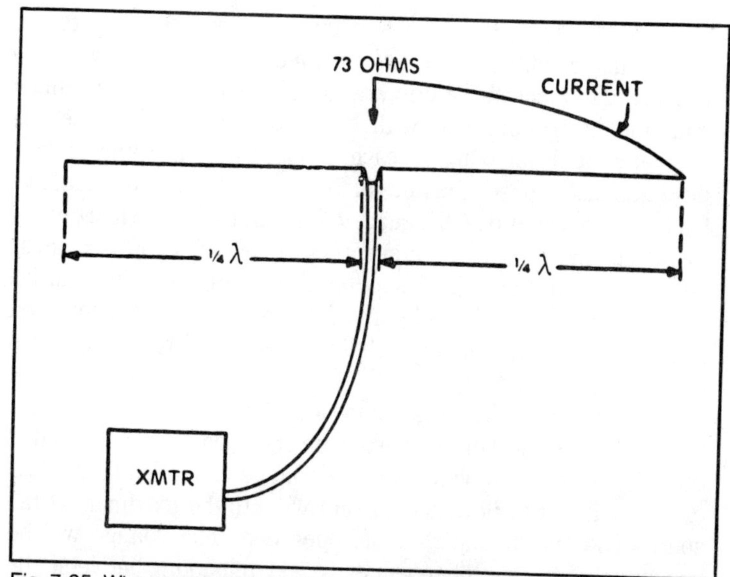

Fig. 7-25. When each quarter-wave element is fed at a high-current point, the voltage at the tip is maximum. If the antenna element is too short, the signal must start traveling back down the antenna to build up to maximum. This "reflected" signal is the source of standing waves.

antenna feed point to tune out the reactance of the transmission line, or to compensate for the inherent reactance of the antenna whose elements are either too short or too long. The tuner effectively changes the length of the antenna elements, which similarly changes the load impedance of the line.

H.2. The Dipole Antenna

(1) Basic characteristics
(2) The image principle
(3) Voltage and current distribution
(4) Length versus frequency

Any antenna having a physical length that is one-half wavelength of the applied frequency is called a *hertz antenna*. Hertz antennas are predominantly used with frequencies above 2 MHz. Usually at frequencies below 2 MHz a *Marconi* type of antenna is used. This is a quarter-wavelength antenna with ground acting as the other quarter wavelength (the *image principle*). Consequently, the difference between the two antennas is that the Hertz type does not require a conducting path to ground while the Marconi type does.

It is desirable to have maximum radiation from an antenna. Under such conditions all energy applied to the antenna would be converted to electromagnetic waves and radiated. This maximum radiation is not possible with the two-wire line because the magnetic field surrounding each conductor of the line is in a direction that opposes the lines of force about the other conductor. Under these conditions, the quarter-wave transmission line proves to be an unsatisfactory antenna; however, with only a slight physical modification, this section of transmission line can be transformed into a relatively efficient antenna. This transmission is accomplished by bending each line outward 90° to form a dipole antenna, as shown in Fig. 7-26.

The antenna shown is composed of two quarter-wave sections. The electrical distance from the end of one line to the end of the other is a half wavelength. If a voltage is applied to the line causing current to flow, the current will still be maximum at the source end and minimum at the open end. The voltage will be maximum between the open ends and minimum between the source ends.

An impedance value may be specified for a half-wave antenna thus constructed. Generally, the impedance at the open ends is maximum while that at the source ends is minimum. Consequently, the impedance value varies from a minimum value at the generator to a maximum value at the open ends. An impedance curve for the half-wave antenna is shown in Fig. 7-27. Notice that the line has different impedance values values for different points along its length. Typical impedance values for half-wave antennas vary from 2500 ohms at the open ends to 73 ohms at the center feed-point.

H.3. Transmission Lines

(1) Types

(a) Parallel conductor

(b) Twisted pair

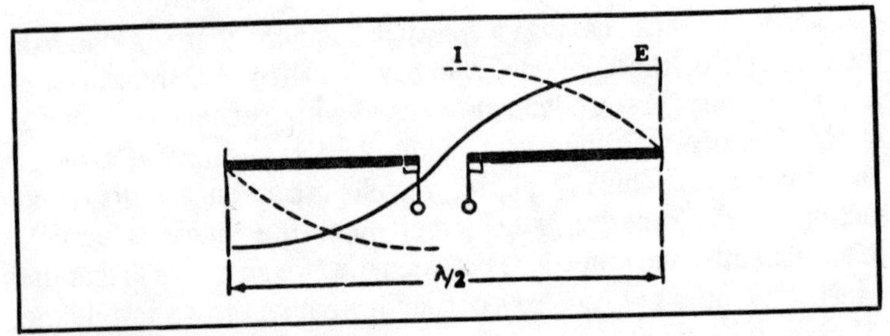

Fig. 7-26. The basic dipole antenna.

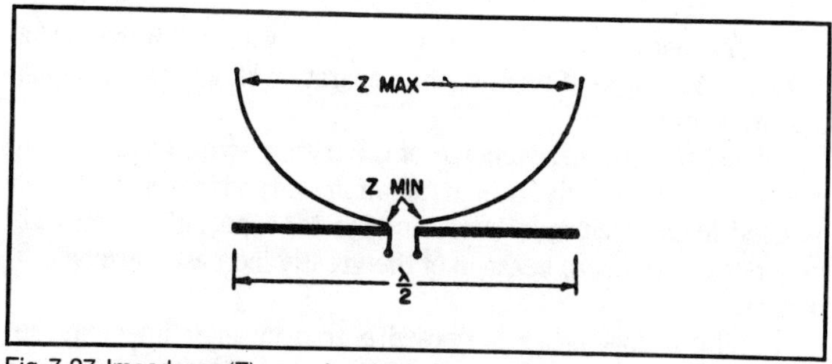

Fig. 7-27. Impedance (Z) curve for a half-wave antenna.

(c) Shielded pair
(d) Coaxial
(2) Standing wave ratio
(3) Matched lines

Parallel Two-Wire Line

One type of transmission line consists of two parallel conductors that are maintained at a fixed distance by means of insulating spacers or spreaders that are placed at suitable intervals. This type of line is shown in Fig. 7-28. The line has the assets of ease of construction, economy, and efficiency. In practical applications two-wire transmission lines (with individual insulators rather than spacers) are used for power lines, rural telephone lines, and telegraph lines. This type of transmission line is also used as the connecting link between an antenna and transmitter or an antenna and receiver.

In practice, such lines used in radio work are generally spaced from 2 to 6 inches apart at frequencies of 14 MHz and below. The maximum spacing for frequencies of 18 MHz and above is 4 inches. In any case, in order to effect the best cancellation of radiation, it is

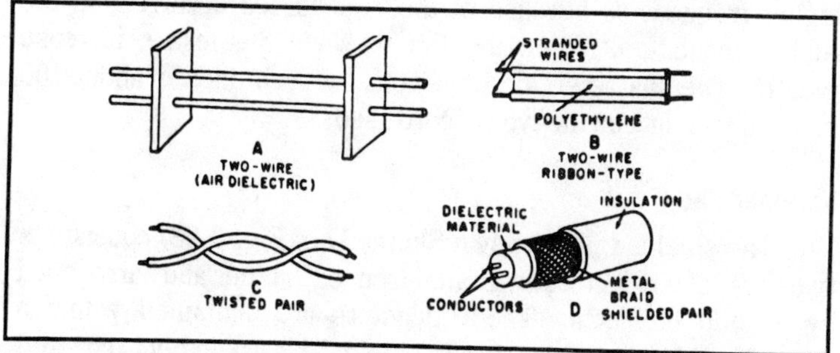

Fig. 7-28. Four types of transmission lines.

necessary that the wires be separated by only a small fraction of a wavelength. For best results, the separation should be less than 0.01 wavelength.

The principal disadvantage of the parallel-wire transmission line is that it has relatively high radiation loss and therefore cannot be used in the vicinity of metallic objects, especially when high frequencies are used, because of the greatly increased loss which results.

Uniform spacing of a two-wire transmission line may be assured if the wires are imbedded in a solid low-loss dielectric throughout the length of the line, as indicated in sketch B. This type of line is often called a two-wire ribbon type. The ribbon type is commonly made with two characteristic impedance values, 300 ohms and 75 ohms. The 300-ohm line is about one-half inch wide and is made of stranded wire. Because the wires are imbedded in only a thin ribbon of polyethylene, the dielectric is partly air and partly polyethylene. Moisture or dirt will change the characteristic impedance of the line. This effect becomes more serious if the line is not terminated in its characteristic impedance.

The wires of the 75-ohm line are closer together, and the field between the wires is confined largely to the dielectric. Weather and dirt affect this line less than they affect the 300-ohm line. The ribbon type of line is widely used to connect television receivers to antennas.

Twisted Pair

The twisted pair is shown in Sketch C of Fig. 7-28. As the name implies, it consists of two insulated wires twisted to form a flexible line without the use of spacers. It is used as an untuned line (on a tuned line the insulation might be broken down by arc-over at voltage loops) for low frequency transmission. It is not used for the higher frequencies because of the high losses occurring in the rubber insulation. When the line is wet, the losses increase greatly. The characteristic impedance of such lines is about 100 ohms, depending on the type of cord used.

Shielded Pair

The shielded pair (shown Sketch D of Fig. 7-28) consists of two parallel conductors separated from each other and surrounded by a solid dielectric. The conductors are contained within a copper-braid tubing that acts as a shield. This assembly is covered with a rubber or flexible composition coating to protect the line against moisture and friction. Outwardly, it looks much like an ordinary power cord for an electric motor.

The principal advantage of the shielded pair is that the two conductors are balanced to ground—that is, the capacitance between each conductor and ground is uniform along the entire length of the line and the wires are shielded against pickup of stray fields. This balance is effected by the grounded shield that surrounds the conductors at a uniform spacing throughout their length.

If radiation from an unshielded line is to be prevented, the current flow in each conductor must be equal in amplitude in order to set up equal and opposite magnetic fields that are thereby canceled out. This condition may be obtained only if the line is clear of all obstructions, and the distance between the wires is small. If, however, the line runs near some grounded or conducting surface, one of the two conductors will be nearer that obstruction than the other. A certain amount of capacitance exists between each of the two conductors and the conducting surface over the length of the line, depending upon the size of the obstruction. This capacitance acts as a parallel conducting between each conductor. Since one conductor may be nearer the obstruction than the other, the current flow will accordingly be increased, resulting in an inequality of current flow in the two conductors and therefore incomplete cancellation of radiation. The shielded pair, therefore, eliminates such losses to a considerable degree by maintaining balanced capacitances to ground.

Coaxial Cable

Coaxial lines, or coaxial cables as they are called, are the most widely used type of rf transmission line. They consist of an outer conductor and an inner conductor held in place exactly at center of the outer conductor.

Several types of coaxial cable have come into wide use for feeding power to an antenna system. Figure 7-29 illustrates the construction of flexible and rigid coaxial cables.

Figure 7-29 shows that one of the conductors is placed inside the other. Since the outside conductor completely shields the inner one, no radiation loss takes place. The conductors may both be tubes, one within the other; the line may consist of a solid wire within a tube, or it may consist of a stranded or solid inner conductor with the outer conductor made of one or two wraps of copper shielding braid.

When using solid dielectric coaxial cables, it is necessary that precautions be taken to insure that moisture cannot enter the line.

If the better grade of connectors manufactured for the line are employed as terminations, this condition is automatically satisfied. If connectors are not used, it is necessary that some type of moisture-proof sealing compound be applied to the end of the cable where it would be exposed to the weather.

The chief advantage of the coaxial line is its ability to keep down radiation losses, very little electric or magnetic fields extend outside the outer conductor. Therefore, nearby metallic objects cause minimum loss, and the coaxial cable may be run up air ducts or elevator shafts, inside bulkheads, or through metal conduit. Insulation troubles can be forgotten. The coaxial cable may be buried in the ground or suspended above ground.

H.4. Sample Question

The approximate length of a half-wave antenna suitable for use by Novice class licensees operating on 40 meters would be:

A. 252 ft.
B. 66 ft.
C. 132 ft.
D. 80 ft.
E. 16 ft.

MISCELLANEOUS TOPICS

You may be asked about the block diagram of a typical radiotelegraphy transmitter, communications receiver, and the layout of a typical Novice-class amateur radio station. In this section, we will discuss these matters in depth sufficient to allow you to pass the examination, plus give you some insight as to the nature of your transmitter.

A block diagram is a circuit-layout diagram that tells us the stages (i.e., types of circuits and their order) needed in a piece of equipment. When we fill in the blocks with the circuitry, therefore, we will have a schematic of the entire equipment.

Figure 7-30 shows the block diagram of a simple radiotelegraph transmitter. Note that the DC power supply, which converts alternating current from the AC power mains to DC through a process of rectification and filtering, is common to all stages except the passive output coupling circuit.

The basic signal is generated in an oscillator stage. The oscillator may be crystal controlled, or it may be a variable frequency oscillator (VFO). The crystal type is more stable, and often more accurate, than the VFO, but limits you to operation on

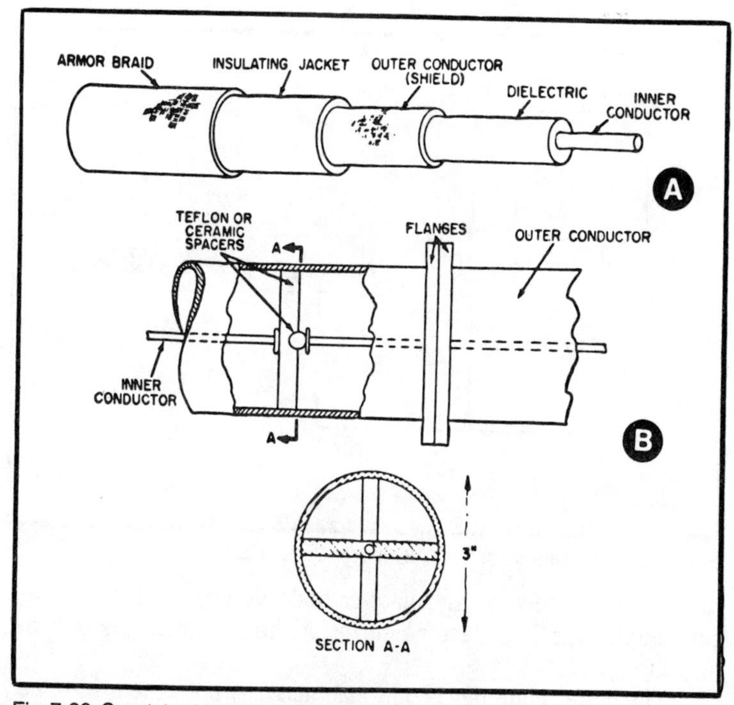

Fig. 7-29. Coaxial cables: (A) Flexible; (B) Rigid.

only the crystal frequency. CB sets are crystal controlled because they have only 40 channels, each a separate frequency. Novice-class amateur radio operators were once restricted to the use of crystal-controlled transmitters because it was felt that their lack of experience made it unlikely that they could properly control a VFO transmitter, and keep it inside of the novice band. But, today, the design of transmitters has improved enough that the FCC now allows novices the use of the VFO and up to 250 watts of power (they had been limited to 75 watts).

Both VFOs and crystal oscillators are operated at low power levels. They are more stable, and will not change frequency as much, when they are not required to produce power sufficient to radiate from the antenna. The power output of the oscillator, then, must be amplified before being applied to the antenna. This is the job of the *final amplifier* stage shown in Fig. 7-30. This stage will have the greatest power consumption. In the novice transmitter, the maximum DC power input to the plate of the vacuum tube, or collector of a transistor, is 250 watts (up to 1000 watts are allowed to amateurs with higher class licenses). We measure this power by

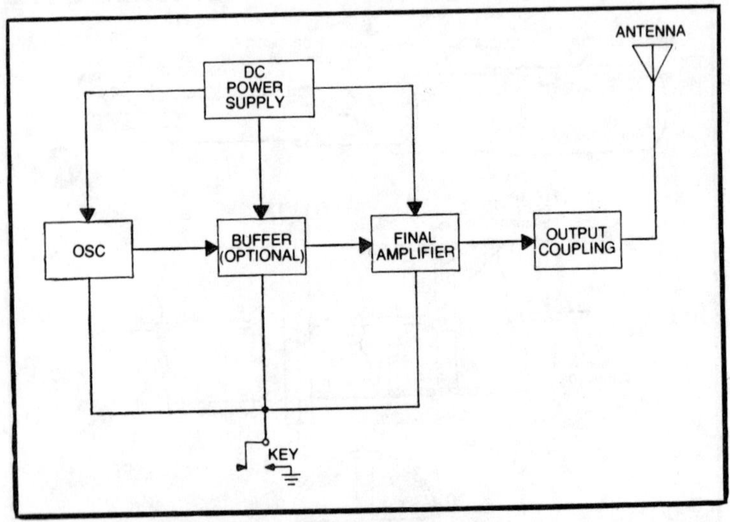

Fig. 7-30. Block diagram of a typical modern transmitter.

taking the product of the plate-cathode voltage and the plate-cathode current (or collector-emitter, in the case of transistor final amplifiers).

The buffer amplifier is optional, and may not appear in some simple radiotelegraphy transmitters. It isolates the oscillator from load variations caused by operation of the final amplifier. Transmitters that lack the buffer are called *master oscillator power amplifier* (MOPA) transmitters, while those that have the buffer are called *master oscillator buffer power amplifier* (MOBPA) transmitters. Simple novice transmitters popular some years ago were almost invariably MOPA designs, while the transceivers used by most novices today are of the MOBPA design.

In some transmitters, the key is used to turn on and off all stages of the transmitter at once. This produces some problems. Chirp is a change of frequency (noticed as a change of pitch by the receiving station) of the oscillator caused by the sudden heavy load placed on a common DC power supply by turning on the power amplifier. There is a type of keying used in some transmitters called *differential keying* in which the final amplifier turn-on is delayed a few milliseconds in order to allow the oscillator to come up to speed before the final amplifier load must be dealt with.

Another problem that is handled by keying is the matter of a *backwave* from the transmitter. The final amplifier is not totally able to suppress the oscillator signal when the amplifier is off. Some small amount of the oscillator signal will still reach the

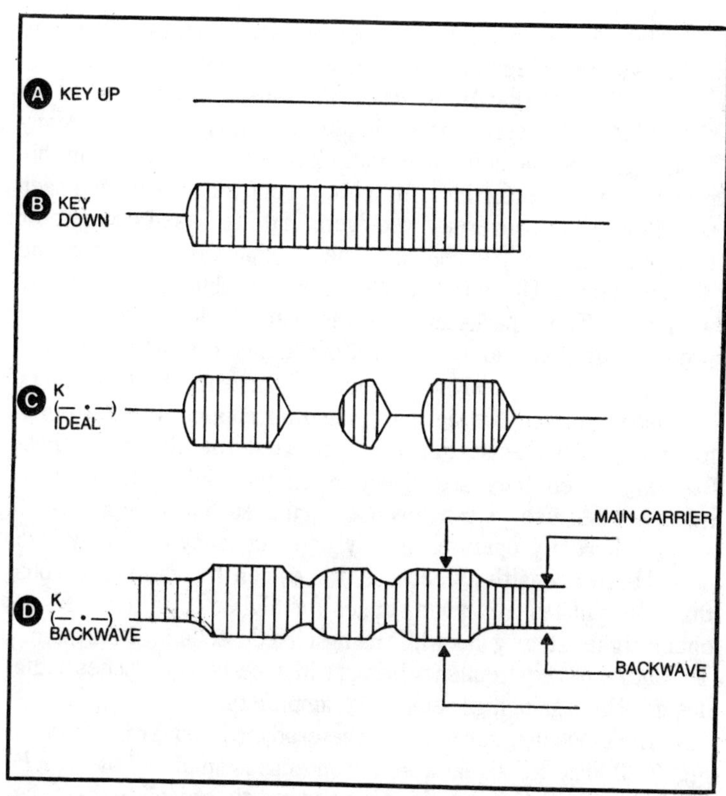

Fig. 7-31. Keyed waveforms at the output of a transmitter.

antenna circuit. In an ideal transmitter the output waveform (Fig. 7-31) will be zero when the key is up and the station is not transmitting (Fig. 7-31A). When the key is held closed, then the transmitter will produce its full RF output until the key is again opened. This produces the constant amplitude shown in Fig. 7-31B. In an ideal transmitter, when a Morse-code letter is transmitted (see Fig. 7-31C for the letter "K"), the amplitude will be zero between characters and maximum during the periods when the key is down. The leading and trailing edges of the waveform are slightly curved by the *key-click filter* used in the keying circuit of the transmitter. When the backwave is present (Fig. 7-31D), the output of the transmitter does not fall all the way back to zero when the key is lifted, but merely drops to some low value. This backwave is transmitted, and may be heard by the receiving station as a constant tone "in the background" while the other station is on the air.

The solution to backwave problems is to turn the oscillator on and off with the telegraph key, along with the amplifier.

Radiotelegraph Receiver. The simplest form of radiotelegraph receiver is a simple oscillating detector. Many years ago, popular shortwave-radio kits were sold based on this regenerative detector idea. In more recent times, we have seen so-called direct conversion receivers that are based on more or less the same idea, except that the oscillator is not part of the detector stage. These radios offer simplicity, but they also fail to work well. The superheterodyne receiver, shown in Fig. 7-32, is probably the most widely used type of communications receiver.

The superheterodyne receiver is superior to the others because it converts the signal from its RF frequency to some other frequency. All frequencies are converted to this *intermediate frequency* when they are tuned in on the dial. Selectivity is improved because we can now use circuits such as crystal filters, which inherently operate on only one frequency. We may also obtain better sensitivity because it is easier, for various reasons that you will learn in your studies for higher class licenses, to obtain higher stage gains while using a fixed frequency I-F system. The ability to vary frequency brings with it certain headaches in the design of high gain, high selectivity, amplifiers.

The block diagram of a superheterodyne receiver is shown in Fig. 7-32. The RF signal from the antenna is amplified by the *RF amplifier* stage. This amplifier is used mostly to isolate the radio from the outside world. Only a small amount of gain is built into the RF amplifier.

The signal from the RF amplifier output is fed to a *mixer* stage, where it is converted to the intermediate frequency. This is the portion of the radio that begets the term "heterodyne." The process of heterodyning can be seen earlier in this chapter where we discussed the BFO in receivers. Remember, mixing together the *A* note, and the *C* note on a piano results in *four* tones (at least): A, C, A−C, and A+C. In the superheterodyne radio receiver, we mix together the RF "tone," and the signal from a Local Oscillator (an LO tone). The output of the mixer will contain the following frequencies: RF, LO, RF+LO, and RF−LO. It has been traditionally the practice to accept only the difference frequency (RF−LO) for the intermediate frequency, but, today, some receivers use the sum tone (RF+LO). Most receivers, however, still find the difference easier to deal with. Popular intermediate frequencies (I1=1F) include 455 kHz, 3385 kHz and 9 MHz. The 455 kHz

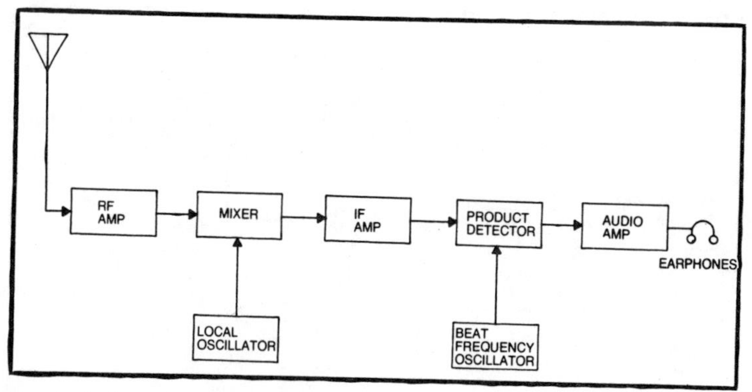

Fig. 7-32. Block diagram of a modern receiver.

frequency is almost universally used for A1=1M broadcast receivers and some amateur receivers of the older and cheaper variety. 3385 kHz was popularized by The Heath Company, while 9 MHz is selected for practical reasons specific to the amateur bands. It is in the *I1=1F amplifier* section of the radio that most of the receiver's gain and selectivity are obtained.

The radiotelegraph signal must be *demodulated* before we can read it. If this process is not done, then the dots and dashes sound like staccato hissing in the output of the receiver. We overcome this problem by using a stage called a *product detector*, which combines the I1=1F signal with that of a *beat frequency oscillator* (BFO). The difference between the I1=1F signal and the BFO signal is an audio note that gives the radiotelegraph signal its characteristic musical quality. The BFO is adjustable in some receivers, but on many modern receivers it is fixed. Most amateurs adjust either the BFO frequency, or the received RF frequency, to cause a beat note in the output of 300 to 1200 hertz (as might be most comfortable for them).

The output of the detector is typically very weak. As a result, an *audio amplifier* is used to amplify the signal to a level that allows it to drive either earphones or loudspeaker.

Station Layout. The equipment in an amateur radio station must be connected in a manner consistent with good operation. It is rarely the case that an amateur will use separate antennas for receiving and transmitting. It is almost universally the case to find a single antenna used for both reception and transmission. The Rule of Reciprocity tells us that the properties of an antenna are the same for receiving as for transmitting, so this makes good sense in most cases.

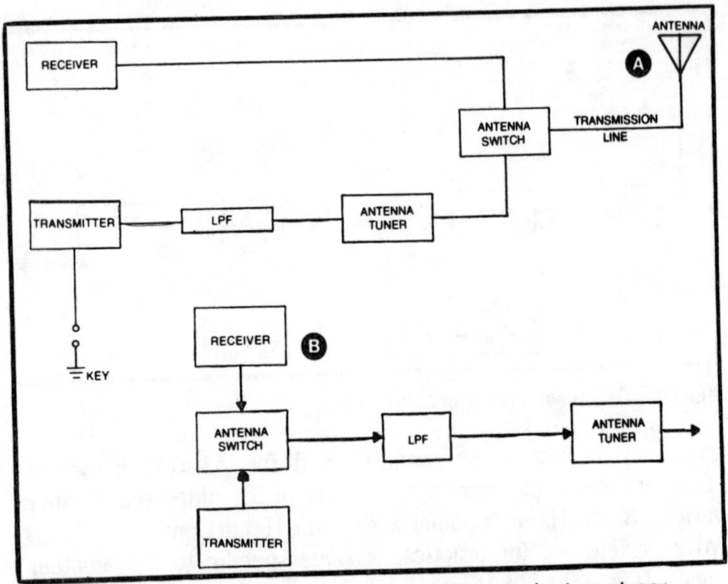

Fig. 7-33. Block diagram showing connections of filters and antenna tuner.

Figure 7-32 shows the layout of a typical amateur radio station. The transmitter is connected to the antenna switch through a *low-pass filter* (to eliminate TVI) and an *antenna tuner* (to impedance-match the antenna, in some cases, and to further reduce TVI in all cases). The receiver and transmitter connect to alternate ports on the antenna switch. This switch will transfer the antenna from the receiver to the transmitter when the operator wants to transmit. In some cases, the antenna switch must be operated manually, or through an external switch. In still other cases, the action is automatic, and the transfer occurs when the electronics inside of the switch sense the presence of RF output from the transmitter. In most cases, a switch or relay inside of the transmitter is operated by the telegraph key, and will turn the antenna switch on for each dot and dash (this is called break-in operation). In some rigs, the manual switch to operate the antenna switch is located on the front panel of the transmitter. Note that transceivers, so popular today, can have any or all of these methods, but they are built into the rig.

An alternative plan is shown in Fig. 7-33A. The difference here is that the low-pass filter and antenna tuner are located in the antenna feedline (transmission line) between the output of the antenna switch and the antenna. For transceivers, this is the usual situation.

Chapter 8
Questions and Answers

The questions in this chapter are representative of those that will be asked on the Novice-class amateur radio examination. These questions are not the *exact* questions that you will see (unless we are better guessers than we think we are!), but cover the same areas. We have worked with FCC personnel on this project, and feel that the subject areas are well covered in this chapter.

The philosophy followed in this chapter is the Q & A format; a question will be asked, then immediately answered. The FCC examination will be multiple choice, so Chapter 9 (Sample Novice Examination) will follow that format. You should feel confident that you will pass the examination if you can get at least 80 percent of the questions in this, and the following, chapter correct without looking. One study aid that at least one author has found useful is to use a 4 × 5 card to cover the answer as you read the question. If you know the answer, verbalize it, and then uncover the printed answer to compare yours with ours. Then move the card down the page to expose the next question. If you get the answer wrong, then place a check mark beside that question and pay special attention to that area in your studies. Read and understand the Regulations in Appendix C.

Q1 *Define amateur radio service*

A. The amateur radio service is defined (97.3a) as a radio communications service of self-training, intercommunication, and technical investigation carried on by amateur radio operators.

Q2 *Define amateur radio station*

A. An amateur radio station (97.3e) is a radio station licensed in the amateur radio service embracing necessary apparatus at a particular location used for amateur radio communication.

Q3 *Define amateur radio operator's license*

A. The instrument of authorization issued by the Federal Communications Commission including the class of operator privileges (97.3d).

Q4 *Define amateur radio station license*

A. The instrument of authorization issued by the F.C.C. for a radio station in the amateur radio service (97.3d).

Q5 *Define control operator*

A. The control operator (97.3o) is a licensed amateur radio operator who is designated by the licensee of an amateur radio station to also be responsible for the emissions from that station.

Q6 *Define amateur radio operator*

A. An amateur radio operator (97.3c) is a person holding a valid license to operate an amateur radio station.

Q7 *Define amateur radio communication*

A. Amateur radio communication (97.3b) is noncommercial radio communication by or among amateur radio stations solely with a personal aim and without pecuniary or business interest.

Q8 *Define third-party traffic*

A. Third-party traffic (97.3v) is amateur radio communication by or under the supervision of the control operator at an amateur radio station to another amateur radio station on behalf of someone other than the control operator.

Note: Other definitions are given in sec. 97.3 of the FCC Rules and Regulations. These should also be studied for sake of familiarity, but those given above are the definitions which the FCC lists in the syllabus for the Novice-class license examination.

Q9 *What types of emission are permitted to the holder of a Novice-class amateur radio license?*

A. On-off radiotelegraphy, also called A1-emission, or A1 radiotelegraphy

Q10 *What frequencies are available to Novice-class operators?*

A.
3700 - 3750 kHz
7100 - 7150 kHz
21,100 - 21,100 kHz
28,100 - 28,200 kHz

Also when a terrestial station operated by a Novice class amateur radio operator is located outside I.T.U. region 2, then the 7050 - 7075 kHz band is also available for novice use. The U.S. is in Region 2. See chart in section 97.75 in Appendix C of this book.

Q11 *May a Novice-class operator use frequencies other than those listed in Q10 when operating a station licensed to an Extra-class operator?*

A. No. The novice operator must use only those frequencies authorized by sec. 97.5e of the FCC Rules and Regulations (see Q 10).

Q12 *May an Extra-class operator use frequencies outside of the novice bands (see Q10 and F.C.C. R&R sec. 97.5e) when operating a station that is licensed to a Novice-class operator?*

A. No. He may use only those frequencies authorized to that station.

Q13 *May you sign your own callsign when operating a station licensed to another amateur?*

A. No. The call sign is issued with the station license, not with the operator's license. You must use the callsign of the station that you are operating.

Q14 *May an operator who holds a Technician, General, Advanced, or Extra-class amateur radio license use single-sideband (SSB) emission when operating a novice station?*

A. No. The amateur operator may use only those types of emission for which the station being operated is licensed. In the case of a novice station, only A1 radiotelegraphy is authorized.

Q15 *Under what circumstances may an amateur radio operator accept payment for transmitting or receiving messages?*

A. An amateur station or operator may not accept payment. The amateur station may not be used to transmit or receive messages for hire, nor for communication for material compensation, direct or indirect, paid or promised.

Note: Control operators of a club station may be compensated for conducting on-the-air radiotelegraph code practice, subject to certain provisions given in sec. 97.112 of the FCC Rules and Regulations.

Q16 *Are there any circumstances under which an amateur operator may transmit a signal under an unauthorized call sign concocted by the operator, or assigned by any person, group, or club, other than the FCC?*

A. No! Only those call signs issued by the FCC are allowed.

Q17 *What is the rule regarding the transmission of false or deceptive signals?*

A. Transmission of false or deceptive signals is strictly prohibited.

Q18 *Under what conditions may an amateur radio operator transmit unidentified signals?*

A. No licensed radio operator shall transmit unidentified radio communications or signals.

Q19 *Is it legal to interfere with another station in order to cause an ill-mannered operator of another station to shut-up?*

A. No. No licensed radio operator shall willfully or maliciously interfere with or cause interference to *any* radio communication or signal . . . no matter what the provocation.

Q20 *Do crowded amateur radio bands mean that you are allowed to "step-on" (i.e., interefere with) another station who is using, or hogging, a frequency that you wish to use?*

A. No. See the answer to Q 19

Q21 *Who issues amateur radio callsigns?*

A. The FCC issues and assigns all callsigns, and will not grant any request for a specific callsign. The FCC occassionally issues Public Notices of changes in callsign policy.

Q22 *What is the basis and purpose of the amateur radio Rules and Regulations.*

A. To recognize and enhance the value of the amateur radio service to the public as a voluntary, non-commercial communication service, particularly with respect to providing emergency communications. 97.1(a)

To continue and extend the amateur radio operators' proven ability to contribute to the advancement of the radio art. 97.1(b)

To encourage and improve the amateur radio service by providing for advancing skills in both the communication and technical phases. 97.1(c)

To expand the existing reservoir within the amateur radio service of trained operators, technicians, and electronics experts. 97.1(d)

To continue and extend the radio amateurs' unique ability to enhance international good will. 97.1(e)

Q23 *What types of radio station may an amateur operator communicate with?*

A. In general, the amateur radio operator may communicate with; a), other amateur radio stations, except in those countries which have notified the ITU that they object to foriegn amateurs communicating with their amateurs; b), stations licensed by the FCC to communicate with amateur stations; c), U.S. Government stations used for Civil Defense purposes, in emergencies or for test purposes; and d), amateur radio stations may also be used for transmitting signals to receivers for the measurement of emissions, temporary observation of radio phenomenon, radio control of remote objects, and for other experimental purposes and those purposes as denoted in sec. 97.91 of the FCC rules and regulations.

Q24 *How long must the log of an amateur station be maintained?*

A. For a period of 1 year following the date of the last entry.

Q25 *What types of data must be entered into the log of an amateur radio station?*

A. The complete answer to this question can be found by reading sec. 97.103 of the FCC Rules and Regulations. But the most important parts are the callsign of the station, the signature of the primary operator (or a photocopy of the operator's license), the locations and dates upon which operation of the station was initiated and terminated, the date and time periods when another operator used the station, and a record of any third-party traffic sent or received, including names and a description of the traffic. See 97.103, Appendix III

Q.26 *How often must an amateur station be identified?*

A. An amateur radio station must be identified (97.84) by the transmission of its call sign at the beginning and end of each transmission, or exchange of transmissions, and at intervals not to exceed 10 minutes during a transmission longer than 10 minutes. Additional requirements for special cases are given in sec. 97.84 of the FCC rules and regulations, see Appendix III).

Q27 *What is the maximum power that can be used by a Novice-class amateur station?*

A. 250 watts dc input to the final stage.

Note: There are some further limitations on power. In general, the FCC requires that the amateur ". . . use the minimum amount of power necessary to carry out the desired communication."

Q28 *How is the power rating of an amateur radio transmitter determined?*

A. The power rating of an amateur radio transmitter is determined to be the dc power applied to the final amplifier, or amplifiers (i.e., the final amplifier(s) is defined as the stage or stages supplying power to the antenna). The final amplifier dc power for vacuum-tube transmitters is the product of the anode (plate) voltage, as measured between the cathode and anode, and the anode current ($P = I_b E_b$). For transistor amplifiers, the dc power is defined as the product of the collector current and the collector-emitter voltage ($P = I_c E_c$).

Note: *This could catch the unwary, because recent definitions includes all dc input. Current flowing in the control grid, the screen grid, the plate, and the cathode, are all considered part of the total dc input. The only current not included is that used to heat the filaments in indirectly heated cathode type tubes.*

Q29 *How long do you have to reply to an official notice of violation from the FCC?*

A. 10 days (see sec. 97.137 for additional details as to completeness of the response and how to handle legitimate delays beyond 10 days).

Q30 *Who is responsible for the operation of an amateur radio station?*

A. The licensee is always responsible for the operation of the station. An amateur station may only be operated to the extent permitted by the operator privileges of the class of license held by the control operator.

Q31 *A signal is described as "599." What does this tell you?*

A. The signal is perfectly readable, is extremely strong, and has a pure dc note.

Q32 *Under what circumstances may a novice operator use radiotelephone emissions such as SSB or FM?*

A. The novice class operator is limited to A1 radiotelegraphy only.

Q33 *What is the proper method for identifying an amateur station when calling another station?*

A. The call sign of the station being called, followed by the letters "DE" (which means "this is"), and then the call letters of the station doing the calling. For example, if K4IPV wishes to call K3RXK, then the proper method would be "K3RXK DE K4IPV."

Q34 *How must the novice station be identified if an Extra-class operator is using the Novice equipment for operation outside of the Novice bands?*

A The proper identification shall be the call sign of the Novice station followed by the Extra-class operator's call sign, with the two separated by a slash bar. For example, if AC4A is operating WN4XYZ, then the proper callsign identification would be WN4XYZ/AC4A.

Q35 *What is the rule regarding third-party traffic with foreign countries?*

A. In general, it is prohibited. There are, however, certain countries which have treaties with the USA that permit third-party traffic. Many of the 20 or so countries which have third-party agreements are in South America and Africa. Since the "third-party traffic list" changes from time to time, we will not print it here. Check with the American Radio Relay League, 225 Main Street, Newington, CT., 06111 for information.

Q36 *What is "zero-beating" of a signal?*

A. In most cases, we want the transmitter frequency to be the same as the frequency of the station being called. If you can hear the other station in the receiver, then turn on the oscillator of the transmitter (not the final amplifier!) and adjust the transmitter frequency until a tone is heard in the receiver. The tone will have a frequency, or pitch, that is equal to the difference between the transmitter frequency and the received frequency. Adjust the transmitter frequency until the tone disappears.

Q37 *Describe briefly the method used for tuning an amateur radio transmitter.*

A. There are two different, but related, methods that depend upon the type of metering used. One case requires the meter to monitor the plate or collector current of the final amplifier, while the other method monitors the relative RF output power. Both methods are described below.

A. When monitoring the plate/anode or collector current:

1. Set the *load* control to its minimum-loading position (if a capacitor, then the plates are fully meshed).

2. Apply a small amount of drive signal to the final amplifier and quickly adjust the *tune* control for a dip in the plate current.
3. Adjust the *load* control for a small increase in plate current, and then dip the plate current again using the *tune* control.
4. Repeat steps 2 and 3 until no further increase is obtained when the loading is increased.
5. Advance the drive-level control until the desired level of plate current is reached.

Note: Many transmitters will have a *grid tuning* control. This control is adjusted initially by monitoring the grid current of the final amplifier, and adjusting the control for a peak reading in grid current.

B. When nonitoring the relative RF output power:
1. Set the *load* control to its minimum position (as in the previous method).
2. Alternately adjust the *tune* and then *load* controls for maximum output, until no further increase is obtained.
3. Advance the drive to the correct output power.

Note: In either method, we will modify the procedure slightly if the transmitter function switch has a *tune* position. In that case, perform all initial tuning in the tune position, and then go to the CW (or TRANS) position for final tuning.

Q38 *How fast should one transmit the radiotelegraph code?*

A. Only fast enough for the other operator to understand. There is a further limitation when operating near the edge of the band, because very high speed keying tends to increase its bandwidth and make you operate slightly outside of the band. This is, however, not usually a problem with novice operators.

Q39 *Give the meaning for the following radiotelegraphy addreviations: CQ, AS, R, DE, K, $\overline{SK}$, $\overline{AR}$, and 73.*

A.
CQ General call to all station.
$\overline{AS}$ Wait, or standby.
R Received correctly all of your transmission. Equivalent to the old term "roger" used on radiotelephone.
DE This is . . .
K Go ahead, or "over"
SK Clear
AR End of transmission

73 Best wishes

Note: Some of these signals, i.e., AS, SK, AR, are sent as one unit rather than two separate letters. For example, SK is sent . . . - . - instead of . . . - . -

Q40 *Give the meaning of the following Q-signals: QRS, QRZ, QTH, QSL, QRM, and QRN.*

A. QRS Shall I send more slowly? Please send more slowly.

QRZ Who is calling me? You are being called by ________.

QTH What is your location? My location is ______.

QSL Acknowledgement

QRM Am I being interfered with? You are being interfered with. (QRM is used for man-made interference by other stations.)

QRN Are you troubled by static? I am being troubled by static. (QRN is similar to QRM, except that static, rather than other stations, is the source of interference.)

Q41 *What is the difference between skywaves and groundwaves?*

A. The groundwave travels along the ground, parallel to the earth's surface, after leaving the transmitting antenna. It is capable of carrying communications for only a few hundred miles in the medium and high frequency ranges, and only a few dozen miles at VHF/UHF frequencies. The skywave leaves the antenna at an angle to the earth's surface, and is reflected back to the earth by the ionosphere. The skywave is responsible for "skip" communications over long distances.

Q42 *Under what circumstances may a private code or cipher be used by an amateur radio station operator?*

A. The amateur operator must use plain language at all times, except for certain internationally recognized telegraphy abbreviations or Q-signals. The use of codes or ciphers, intended to conceal the meaning of the transmission, is prohibited.

Q43 *What provisions must be made to prevent unauthorized persons from using an amateur radio station?*

A. The licensee shall make whatever provisions are needed to prevent such operation of the amateur radio station.

Q44 *What are some of the precautions needed to protect an amateur station from lightning damage?*

A. Use a grounded feedline. If a crank-up tower is used, then lower it to the lowest height possible. Make sure all equipment is grounded to a ground rod in the earth. Disconnect the radio equipment from both the antenna feedline and the AC power mains.

Q45 *Why is a good ground system required in an amateur radio station?*

A. To ensure proper operation of the equipment and antenna, and to provide some measure of safety for the operator. A good ground is one that has a long conductive rod into the earth, and a short (relative to one wavelength) heavy conductor to the transmitter. The importance of a good ground is difficult to overestimate.

Q46 *What are some safety precautions to observe when installing an antenna?*

A. Besides the common-sense precautions (don't fall off the ladder), it is *essential* that you avoid the power lines. Don't install the antenna where it can fall over and contact the power lines either during installation or later. This is another precaution that is difficult to overstate... those power lines are lethal. They may appear safe because of the insulation, but such insulation is often rotted and will expose the wire if touched. Several amateur deaths are reported each year after someone tries to install a wire antenna by throwing the wire over a power line! Do not have the antenna feedline connected to the transmitter while installing an antenna.

Q47 *What is TVI? BCI?*

A. TVI refers to television interference, i.e., interference by an amateur station to television receivers. BCI is the equivalent for broadcast radio receiver interference.

Q48 *What steps can be taken if it is found that TVI is caused by spurious emissions from a radio transmitter?*

A. Provide good grounding for the transmitter and antenna, use a single-band antenna, provide a low-pass filter between the output of the transmitter and the antenna, use an antenna tuner or coupler. In practice, many amateurs use all of these methods in order to ensure harmonic and parasitic-free operation of their equipment.

Q49 *What steps can be taken if it is found that TVI is caused by simple overload of the TV receiver?*

A. Reduce operating power, have the TV owner install a high-pass filter on the TV.

Q50 *What are "acceptable" SWR readings on an antenna system?*

A. The lower the better. SWR, also called VSWR, is given as a ratio, such as 2:1. The meaning of SWR is somewhat clouded by a lot of jargon and rhetoric, but in general, the lower the better. Anything under 2:1 is acceptable in most cases.

Q51 *What could be a possible cause of a high SWR reading?*

A. Defects in the antenna, such as open conductor, radiator touching a tree, etc. Defects (shorts or opens) in the transmission line. Improper tuning of the antenna. The length of the transmission line, despite what some in CB circles contend, has no effect on the SWR of an antenna system. This only *appears* to be true due to deficiencies in the measurement system used.

Q52 *What is "voltage?"*

A. Voltage is the potential, or pressure, in an electrical circuit. The voltage is analogous to the pressure in a plumbing system, and causes the current to flow.

Q53 *Define conductors and insulators.*

A. A conductor has lots of free electrons, so will permit the passage of an electrical current. An insulator has few free electrons, so will not easily pass an electrical current.

Q54 *Give several examples of conductors.*

A. Most metals (copper, silver, gold, etc.), acids, organic materials, and wet materials. (*Pure* water is an insulator, but water found in most places contains impurities that makes it conduct electricity. Tap and well water is an excellent conductor, so care must be used when working on electrical devices in a wet environment.)

Q55 *Give several examples of insulators.*

A. Dry wood, glass, ceramic, bakelite, rubber, most plastics.

Q56 *What is meant by "RF?"*

A. RF means radio frequency, i.e., those frequencies of alternating current that can be used for radio communications. Generally, all frequencies above 20 kHz or so (some authorities claim 100 kHz).

Q57 *What is power?*

A. It is a measure of the work that can be done in an electrical circuit. For most purposes, the power is the product of the current and voltage.

Q58 *What is a direct current? An alternating current?*

A. A direct current is the type of electrical current obtained from batteries and certain other sources. It flows in only one direction, and has a constant value. Alternating current, on the other hand, reverses direction periodically. The frequency of the alternating current is a measure of the numbers of times per second that these reversals take place. The AC will begin at zero, rise to a positive maximum, and then drop back to zero. It then reverses direction, and repeats the process. Each cycle consists of the flow in both directions, i.e., one alternation in each of the two directions.

Q59 *What is a short circuit?*

A. A short circuit is one which contains an unwanted path for electrical current that bypasses the regular circuit.

Q60 *What is an open circuit?*

A. An open circuit exists when there is no path for current to flow. This may be intentional, as in the case of an open switch, or accidental.

Q61 *What is the relationship between the frequency and the wavelength of a radio signal?*

A. The frequency is the number of alternations of the AC signal per unit of time. A 7100 kHz signal, for example, will have a frequency of 7,100,000 alternations per second. The wavelength of the signal is the physical length in free-space of *each cycle*. The wavelength is usually measured in meters, while the frequency is measured in hertz (cycles per second).

Mathematically:

$$\text{Wavelength (meters)} = \frac{300{,}000{,}000}{\text{Frequency(Hz)}}$$

When kilohertz or megahertz are used, it can be written:

$$\lambda\text{meters} = \frac{300{,}000}{F_{kHz}}$$

or:

$$\lambda meters = \frac{300}{F_{MHz}}$$

The symbol "λ" (Greek "lambda") is often used to represent wavelength. The term 300,000,000 is the velocity of light.

Q62 *What are the audio frequencies?*

A. The audio frequencies are those AC frequencies within the range of human hearing, generally, 20 to 20,000 hertz (although it takes a good ear to hear either the high or low frequency extremes!).

Q63 *What are the units of electrical potential, or "electromotive force" (EMF)?*

A. Volts.

Q64 *What are the units of electrical current?*

A. Amperes.

Q65 *What are the units of electrical power?*

A. Watts.

Q66 *What are the units of AC frequency?*

A. Hertz (or cycles per second).

Q67 *What is the relationship between hertz and cycles-per-second?*

A. One hertz is one cycle-per-second (1 Hz = 1 cps).

Q68 *What are the units of inductance?*

A. Henrys.

Q69 *What are the units of capacitance?*

A. Farads.

Q70 *What are the commonly used metric prefixes for electrical units, and how are they applied?*

A. Mega = 1,000,000
Kilo = 1,000
Centi = 0.01 (1/100)
Milli = 0.001 (1/1000)
Micro = 0.000001 (1/1,000,000)

The metric prefixes are appended to the basic units to denote higher or lower orders of the same unit. For example, if 1 Hz is 1 cps, then 1 megahertz (1 MHz) is 1,000,000 cycles per second. If we want to use lower numbers to represent an electromotive force of 5000 volts, we might write 5 kV (5 kilovolts, or 5 thousands of volts). Similarly, we find many practical electrical quantities too small for use of the basic unit. An example is the input signal to a radio receiver is microvolts, NV (1/1,000,000 volt), while the plate current of a popular 100-watt transmitter is 150 milliamperes (mA).

Q71 *How many amperes are 400 mA (milliamperes)?*

A. 400 mA = 0.4 A.

Q72 *What is the physical length required of a quarter-wavelength vertical antenna that is to be operated on a frequency of 21,000 kHz?*

A. 11.1 feet. The formula for a quarter wavelength antenna (practical antenna) is: Length (ft.) = 234/Frequency (MHz), so

$$L = 234/21.1 = 11.1 \text{ feet}$$

Q73 *What is the physical length of a half wavelength dipole antenna intended for operation on a frequency of 7.15 MHz?*

A. 65.5 feet. ($L = 468/F = 468/7.15 = 65.5$ feet)

Q74 *What steps can be taken to prevent undesirable harmonic radiation?*

A. See question Q48.

Q75 *What is a possible cause of chirp in a radiotelegraph signal?*

A. Chirp is an emission defect that is caused by a slight shifting of the transmitter frequency when the key is depressed. It can be caused by poor isolation between the oscillator and its load, improper filtering (or defective filtering) of the DC power supply, a defective oscillator, improper regulation of the oscillator DC voltage supply, or a defective crystal.

Q76 *What is a backwave and how may it be prevented?*

A. A backwave is a signal emitted from your transmitter when the telegraph key is up. It is caused by the oscillator signal being coupled through the final amplifier to the antenna. It can be eliminated by turning the oscillator on and off with the telegraph key, along with the final amplifier.

Q77 *What can be done to reduce key clicks from a radiotelegraph transmitter?*

A. Install a keyclick filter between the key and the transmitter.

Q78 *What can be a possible cause of hum superimposed on an amateur signal?*

A. In radiotelegraph transmitters the principal cause is poor, or defective, filtering of the DC power supply. Replacement of the filter capacitors with known good capacitors, or increasing the value of filters, can help eliminate superimposed hum. The hum comes from ineffective reduction of the ripple component of the DC caused by the rectifiers.

Q79 *What can be done about spurious emissions from a transmitter?*

A. If the emissions are parastics, then a parasitic choke in the plate lead of the final amplifier tubes can be used. Otherwise, use the techniques of Q48 (TVI question).

Q80 *What is A1 emission?*

A. On-off radiotelegraphy in which a telegraph key is used to turn the transmitter carrier signal on and off in accordance with the International Morse code. A1 is also sometimes called CW for "continuous wave."

Q81 *What is the purpose of a fuse?*

A. A fuse protects the equipment from overload and short circuits. A general rule of thumb is "fuses don't *cause* trouble, they *indicate* trouble!"

Chapter 9
A Sample Novice Exam

ELEMENT 2

1. The unit of electrical resistance is the
   - (a) ohm
   - (b) watt
   - (c) gilbert
   - (d) joule
   - (e) ampere
2. The unit of electromotive force is the
   - (a) ohm
   - (b) mho
   - (c) volt
   - (d) maxwell
   - (e) ampere
3. The joule is the unit of electrical
   - (a) power
   - (b) resistance
   - (c) reluctance
   - (d) admittance
   - (e) energy
4. Current is to ampere as power is to
   - (a) gilbert
   - (b) oersted
   - (c) ampere-hour
   - (d) joule
   - (e) watt

5. Which of the following is not one of the purposes of the amateur radio service?

(a) To stimulate the interchange of creative ideas among citizens of the U.S.
(b) To maintain a voluntary noncommercial communications service.
(c) To advance the state of the radio art.
(d) Creation of a reservoir of trained radio operators and electronics experts.
(e) To enhance international goodwill.

6. The maximum dc power input to the final of a Novice transmitter is

(a) 75 ampere-hours
(b) 250 watts
(c) 100 ampere-hours
(d) 100 watts
(e) None of the above.

7. Novice licensees may use radiotelephony on which of the following bands

(a) 75 meters
(b) 40 meters
(c) 10 meters
(d) 2 meters
(e) None of the above.

8. Broadcasting is permitted

(a) when the audience is composed only of Novice licensees.
(b) when the content of your message is of direct interest to all listeners
(c) when no individual amateur answers your initial call
(d) never.
(e) When transmissions do not exceed 10 minutes in length.

9. A Novice license is issued for

(a) one year, nonrenewable.
(b) one year, renewable.
(c) two years, nonrenewable.
(d) five years, renewable.
(e) None of the above.

10. Sky waves and ground waves are

(a) always in phase.
(b) never in phase.
(c) horizontally polarized.

(d) propagated from antennas.
(e) a form of mutual inductance.

11. Radio waves travel at what speed in free space?
(a) 300 meters per second
(b) 300 million meters per second
(c) 300 meters per minute
(d) 186,000 miles per hour
(e) None of the above.

12. The electrical length of a wave
(a) depends on the medium through which the wave is propagated.
(b) is equal to the speed of light.
(c) is always equal to the wave's physical length.
(d) is equal to the square root of the physical length divided by its LC ratio.
(e) is an unvarying constant.

13. Sunspots tend to occur
(a) daily.
(b) in 11-year cycles.
(c) during periods of full moon.
(d) on overcast summer days.
(e) only during winter.

14. To receive a signal from a horizontally polarized antenna, the receiving antenna should ideally be polarized
(a) horizontally.
(b) vertically.
(c) in a cross vector with respect to the transmitter.
(d) at 73 ohms.
(e) None of the above.

15. The formula for wavelength is
(a) f/c
(b) c/f
(c) I^2R
(d) E/R
(e) frequency squared times electrical length.

16. The term QSY without a question mark means
(a) Your code speed is too fast; please send slower.
(b) Please wait a few moments before responding.
(c) Please change frequency.
(d) I am having trouble reading you.
(e) None of the above.

17. Spurious radiation should be
   (a) avoided.
   (b) optimized.
   (c) tuned to a peak on the final plate current meter.
   (d) tuned to a dip as measured on the final plate current meter.
   (e) enhanced with positive feedback.
18. The symbol A0 means
   (a) pure carrier with key held down.
   (b) transmitted intelligence with a telegraph key.
   (c) amplitude-modulated telephony.
   (d) All of the above.
   (e) None of the above.
19. Ohm's law states that
   (a) $R = EI$
   (b) $E = I/R$
   (c) $I = ER$
   (d) $E = IR$
   (e) None of the above.
20. The unit of capacitance is the
   (a) capacitor
   (b) ohm
   (c) farad
   (d) micromho
   (d) gilbert
21. A bipolar transistor may be used as
   (a) an amplifier
   (b) an oscillator
   (c) a common-collector stage
   (d) a common-emitter stage
   (e) All of the above.
22. An emitter follower is the transistor counterpart to a vacuum-tube
   (a) grounded-grid circuit
   (b) grounded-cathode circuit
   (c) common-plate circuit.
   (d) common-drain circuit.
   (e) None of the above.
23. The ideal standing-wave ratio on a transmission line is
   (a) 1.414 to 1.
   (b) 2.28 to 1.
   (c) equal to the turns ratio of the link coupling.

(d) 1 to 1.
(e) None of the above.

24. The unit of electrical frequency is the
(a) henry
(b) faraday
(c) voltampere
(d) hertz
(e) gauss

25. The total current through a line in which two paralleled 20-ohm resistors are placed in series with a 12V battery would be
(a) 1.2 amperes.
(b) 0.3 amperes.
(c) 120 amperes.
(d) 2.4 amperes.
(e) None of the above.

ANSWERS

1(a)	4(e)	7(e)	10(d)	13(b)	16(c)	19(d)	22(c)	25(a)
2(c)	5(a)	8(d)	11(b)	14(a)	17(a)	20(c)	23(d)	
3(e)	6(b)	9(d)	12(a)	15(b)	18(a)	21(e)	24(d)	

Appendix A
Novice Study Guide:
The FCC Syllabus

STUDY GUIDE FOR THE AMATEUR RADIO OPERATOR LICENSE EXAMINATIONS

This Bulletin contains syllabi for the FCC amateur radio examinations.

Why Are Amateur Radio Operator Examinations Required?

The examinations determine if you are qualified for the privileges conveyed by an amateur radio license. Those privileges are many and diverse. As an amateur radio operator, you will be allowed to build, repair, and modify your radio transmitters. You will be responsible for the technical quality of your station's transmissions. You will be allowed to communicate with amateur radio operators in other countries around the world and, in some cases, send messages for friends. As you upgrade to the higher operator license classes, you will be allowed to communicate using not only telegraphy and voice, but also teleprinting, facsimile, and several forms of television. For such a flexible radio service to be practical, you and every other amateur radio operator must thoroughly understand your responsibilities and develop the skills needed to operate your amateur radio station properly.

What Subjects Do The Amateur Radio Examinations Cover?

The examinations cover the rules, practices, procedures, and technical material that you will need to know in order to operate your amateur radio station properly. Each examination element is composed of questions which will determine whether you have an adequate understanding of the topics listed in the corresponding syllabus. For example, all Element 3 examination questions are derived from the Element 3 syllabus, which appears on pages 5, 6, and 7 of this Bulletin. To properly prepare for an examination, you should become knowledgeable about all of the topics in the syllabus for the lement you will be taking. Every examination covers nine general subjects:

Rules and Regulations
Circuit Components
Antennas and Feedlines
Electrical Principles
Practical Circuits
Radio Wave Propagation
Signals and Emissions
Operating Procedures
Amateur Radio Practice

Periodically, the syllabi are updated to reflect changing technology and amateur radio practices. Comments on the study guide contents are welcome. Mail them to:

Personal Radio Branch
Federal Communications Commission
Washington, D.C. 20554

Where Can Study Manuals Be Obtained?

A study manual can be helpful in preparing for an examination. Several publishers offer manuals or courses based upon the material in this Bulletin. These may be found in many public libraries and radio stores. The FCC does not offer such manuals, nor recommend any specific publisher. However, you will find two FCC publications, *Part 97 - Rules and Regulations for the Amateur Radio Service* and *How to Identify and Resolve Radio-TV Interference Problems*, useful when preparing for the amateur radio examinations. Copies are sold by the Superintendent of Documents, U.S. Government Printing Office, Washington D.C. 20402. Specify stock number 004-000-00345-4 for the Radio-TV interference booklet.

Study Topics For The Novice Class Amateur Radio Operator License Examination

(Element 2 Syllabus)

A. RULES AND REGULATIONS

DEFINE:

(1) AMATEUR RADIO SERVICE 97.3(a)
(2) AMATEUR RADIO OPERATOR 97.3(c)
(3) AMATEUR RADIO STATION 97.3(e)
(4) AMATEUR RADIO COMMUNICATIONS 97.3(b)
(5) OPERATOR LICENSE 97.3(d)
(6) STATION LICENSE 97.3(d)
(7) CONTROL OPERATOR 97.3(o)
(8) THIRD PARTY TRAFFIC 97.3(v)

NOVICE CLASS OPERATOR PRIVILEGES:

(9) AUTHORIZED FREQUENCY BANDS 97.7 (e)
(10) AUTHORIZED EMISSION (A1) 97.7(e)

PROHIBITED PRACTICES:

(11) UNIDENTIFIED COMMUNICATIONS 97.123
(12) INTENTIONAL INTERFERENCE 97.125
(13) FALSE SIGNALS 97.121
(14) COMMUNICATION FOR HIRE 97.112(a)

BASIS AND PURPOSE OF THE AMATEUR RADIO SERVICE RULES AND REGULATIONS:

(15) TO RECOGNIZE AND ENHANCE THE VALUE OF THE AMATEUR RADIO SERMICE TO THE PUBLIC AS A VOLUNTARY, NON-COMMERCIAL COMMUNICATION SERVICE, PARTICULARLY WITH RESPECT TO PROVIDING EMERGENCY COMMUNICATIONS. 97.1(a)
(16) TO CONTINUE AND EXTEND THE AMATEUR RADIO OPERATORS' PROVEN ABILITY TO CONTRIBUTE TO THE ADVANCEMENT OF THE RADIO ART. 97.1(b)
(17) TO ENCOURAGE AND IMPROVE THE AMATEUR RADIO SERVICE BY PROVIDING FOR ADVANCING SKILLS IN BOTH THE COMMUNICATION AND TECHNICAL PHASES. 97.1(c)

(18) TO EXPAND THE EXISTING RESERVOIR WITHIN THE AMATEUR RADIO SERVICE OF TRAINED OPERATORS, TECHNICIANS, AND ELECTRONICS EXPERTS. 97.1(d)

(19) TO CONTINUE AND EXTEND THE RADIO AMATEURS' UNIQUE ABILITY TO ENHANCE INTERNATIONAL GOOD WILL. 97.1(e)

OPERATING RULES:

(20) U.S. AMATEUR RADIO STATION CALL SIGNS 2.302 and FCC Public Notice

(21) PERMISSIBLE POINTS OF COMMUNICATIONS 97.89(a)(1)

(22) STATION LOGBOOK, LOGGING REQUIREMENTS 97.103(a), (b); 97.105

(23) STATION IDENTIFICATION 97.84(a)

(24) NOVICE BAND TRANSMITTER POWER LIMITATION 97.67(b), (d)

(25) NECESSARY PROCEDURE IN RESPONSE TO AN OFFICIAL NOTICE OF VIOLATION 97.137

(26) CONTROL OPERATOR REQUIREMENTS 97.79(a), (b)

B. OPERATING PROCEDURES

(1) R-S-T SIGNAL REPORTING SYSTEM

(2) CHOICE OF TELEGRAPHY SPEED

(3) ZERO-BEATING RECEIVED SIGNAL

(4) TRANSMITTER TUNE-UP PROCEDURE

(5) USE OF COMMON AND INTERNATIONALLY RECOGNIZED TELEGRAPHY ABBREVIATIONS, INCLUDING: CQ, DE, K, SK, R, $\overline{\text{AR}}$, 73, QRS, QRZ, QTH, QSL, QRM, QRN

C. RADIO WAVE PROPAGATION

(1) SKY WAVE; "SKIP" (2) GROUND WAVE

D. AMATEUR RADIO PRACTICE

(1) MEASURES TO PREVENT USE OF AMATEUR RADIO STATION EQUIPMENT BY UNAUTHORIZED PERSONS

SAFETY PRECAUTIONS:

(2) LIGHTNING PROTECTION FOR ANTENNA SYSTEM

(3) GROUND SYSTEM

(4) ANTENNA INSTALLATION SAFETY PROCEDURES

ELECTROMAGNETIC COMPATABILITY IDENTIFY AND SUGGEST CURE:

(5) OVERLOAD OF CONSUMER ELECTRONIC PRODUCTS BY STRONG RADIO FREQUENCY FIELDS
(6) INTERFERENCE TO CONSUMER ELECTRONIC PRODUCTS CAUSED BY RADIATED HARMONICS

INTERPRETATION OF S.W.R. READINGS AS RELATED TO FAULTS IN ANTENNA SYSTEM:

(7) ACCEPTABLE READINGS
(8) POSSIBLE CAUSES OF UNACCEPTABLE READINGS

E. ELECTRICAL PRINCIPLES

CONCEPTS:

(1) VOLTAGE
(2) ALTERNATING CURRENT, DIRECT CURRENT
(3) CONDUCTOR, INSULATOR
(4) OPEN CIRCUIT, SHORT CIRCUIT
(5) ENERGY, POWER
(6) FREQUENCY, WAVELENGTH
(7) RADIO FREQUENCY
(8) AUDIO FREQUENCY

ELECTRICAL UNITS:

(9) VOLT
(10) AMPERE
(11) WATT
(12) HERTZ
(13) METRIC PREFIXES: MEGA, KILO, CENTI, MILLI, MICRO, PICO

F. CIRCUIT COMPONENTS

PHYSICAL APPEARANCE, APPLICATIONS, AND SCHEMATIC SYMBOLS OF:

(1) QUARTZ CRYSTALS
(2) METERS (D'ARSONVAL MOVEMENT)
(3) VACUUM TUBES
(4) FUSES

G. PRACTICAL CIRCUITS

BLOCK DIAGRAMS:

(1) THE STAGES IN A SIMPLE TELEGRAPHY (A1) TRANSMITTER
(2) THE STAGES IN A SIMPLE RECEIVER CAPABLE OF TELEGRAPHY (A1) RECEPTION
(3) THE FUNCTIONAL LAYOUT OF NOVICE STATION EQUIPMENT, INCLUDING TRANSMITTER, RECEIVER, ANTENNA SWITCHING, ANTENNA FEEDLINE, ANTENNA, AND TELEGRAPH KEY

H. SIGNALS AND EMISSIONS

(1) EMISSION TYPE A1

CAUSE AND CURE:

(2) BACKWAVE
(3) KEY CLICKS
(4) CHIRP
(5) SUPERIMPOSED HUM
(6) UNDESIRABLE HARMONIC EMISSIONS
(7) SPURIOUS EMISSIONS,

I. ANTENNAS AND FEEDLINES

NECESSARY PHYSICAL DIMENSIONS OF THESE POPULAR HIGH FREQUENCY ANTENNAS FOR RESONANCE ON AMATEUR RADIO FREQUENCIES:

(1) A HALF-WAVE DIPOLE
(2) A QUARTER-WAVE VERTICAL

COMMON TYPES OF FEEDLINES USED AT AMATEUR RADIO STATIONS

(3) COAXIAL CABLE
(4) PARALLEL CONDUCTOR LINE

Appendix B
General/Technician
Study Guide: The FCC Syllabus

STUDY TOPICS FOR THE TECHNICIAN/GENERAL CLASS AMATEUR RADIO OPERATOR LICENSE EXAMINATION

(Element 3 Syllabus)

A. RULES AND REGULATIONS

(1) CONTROL POINT 97.3(p)
(2) EMERGENCY COMMUNICATIONS 97.3(w); 97.107
(3) AMATEUR RADIO TRANSMITTER POWER LIMITATIONS 97.67
(4) STATION IDENTIFICATION REQUIREMENTS 97.84(b), (f), (g); 97.79(c)
(5) THIRD PARTY PARTICIPATION IN AMATEUR RADIO COMMUNICATIONS 97.79(d)
(6) DOMESTIC AND INTERNATIONAL THIRD PARTY TRAFFIC 97.114; Appendix 2, Art. 41, Sec. 2
(7) PERMISSIBLE ONE-WAY TRANSMISSIONS 97.91
(8) FREQUENCY BANDS AVAILABLE TO THE TECHNICIAN CLASS 97.7(d)
(9) FREQUENCY BANDS AVAILABLE TO THE GENERAL CLASS 97.7(b)
(10) LIMITATIONS ON USE OF AMATEUR RADIO FREQUENCIES 97.61
(11) SELECTION AND USE OF FREQUENCIES 97.63
(12) RADIO CONTROLLED MODEL CRAFTS AND VEHICLES 97.65(a); 97.99
(13) RADIOTELEPRINTER EMISSIONS 97.69
PROHIBITED PRACTICES:

(14) BROADCASTING 97.113
(15) MUSIC 97.115
(16) CODES AND CIPHERS 97.117
(17) OBSCENITY, INDECENCY, PROFANITY 97.119

B. OPERATING PROCEDURES

(1) RADIOTELEPHONY
(2) RADIO TELEPRINTING
(3) USE OF REPEATERS
(4) VOX TRANSMITTER CONTROL
(5) FULL BREAK-IN TELEGRAPHY
(6) OPERATING COURTESY
(7) ANTENNA ORIENTATION
(8) INTERNATIONAL COMMUNICATION
(9) EMERGENCY PREPAREDNESS DRILLS

C. RADIO WAVE PROPAGATION

(1) IONOSPHERIC LAYERS; D, E, F1, F2
(2) ABSORPTION
(3) MAXIMUM USABLE FREQUENCY
(4) REGULAR DAILY VARIATIONS
(5) SUDDEN IONOSPHERIC DISTURBANCE
(6) SCATTER
(7) SUNSPOT CYCLE
(8) LINE-OF-SIGHT
(9) DUCTING, TROPOSPHERIC BENDING

D. AMATEUR RADIO PRACTICE

SAFETY PRECAUTIONS:

(1) HOUSEHOLD AC SUPPLY AND ELECTRICAL WIRING SAFETY
(2) DANGEROUS VOLTAGES IN EQUIPMENT MADE INACCESSIBLE TO ACCIDENTAL CONTACT

TRANSMITTER PERFORMANCE:

(3) TWO TONE TEST
(4) NEUTRALIZING FINAL AMPLIFIER
(5) POWER MEASUREMENT

USE OF TEST EQUIPMENT:

(6) OSCILLOSCOPE
(7) MULTIMETER
(8) SIGNAL GENERATORS
(9) SIGNAL TRACER

ELECTROMAGNETIC COMPATIBILITY; IDENTIFY AND SUGGEST CURE:

(10) DISTURBANCE IN CONSUMER ELECTRONIC PRODUCTS CAUSED BY AUDIO RECTIFICATION

PROPER USE OF THE FOLLOWING STATION COMPONENTS AND ACCESSORIES:

(11) REFLECTOMETER (VSWR METER)

(12) SPEECH PROCESSOR - RF AND AF

(13) ELECTRONIC T-R SWITCH

(14) ANTENNA TUNING UNIT; MATCHING NETWORK

(15) MONITORING OSCILLOSCOPE

(16) NON-RADIATING LOAD; "DUMMY ANTENNA"

(17) FIELD STRENGTH METER; S-METER

(18) WATTMETER

E. ELECTRICAL PRINCIPLES

CONCEPTS:

(1) IMPEDANCE
(2) RESISTANCE
(3) REACTANCE
(4) INDUCTANCE
(5) CAPACITANCE
(6) IMPEDANCE MATCHING

ELECTRICAL UNITS:

(7) OHM
(8) MICROFARAD, PICOFARAD
(9) HENRY, MILLIHENRY, MICROHENRY
(10) DECIBEL

MATHEMATICAL RELATIONSHIPS:

(11) OHM'S LAW

(12) CURRENT AND VOLTAGE DIVIDERS

(13) ELECTRICAL POWER CALCULATIONS

(14) SERIES AND PARALLEL COMBINATIONS; OF RESISTORS, OF CAPACITORS, OF INDUCTORS

(15) TURNS RATIO; VOLTAGE, CURRENT, AND IMPEDANCE TRANSFORMATION

(16) ROOT MEAN SQUARE VALUE OF A SINE WAVE ALTERNATING CURRENT

F. CIRCUIT COMPONENTS

PHYSICAL APPEARANCE, TYPES, CHARACTERISTICS, APPLICATIONS, AND SCHEMATIC SYMBOLS FOR:

(1) RESISTORS
(2) CAPACITORS
(3) INDUCTORS
(4) TRANSFORMERS
(5) POWER SUPPLY TYPE DIODE RECTIFIERS

G. PRACTICAL CIRCUITS

(1) POWER SUPPLIES
(2) HIGH-PASS, LOW-PASS, AND BAND-PASS FILTERS
(3) BLOCK DIAGRAMS SHOWING THE STAGES IN COMPLETE AM, SSB, AND FM TRANSMITTERS AND RECEIVERS

H. SIGNALS AND EMISSIONS

(1) EMISSION TYPES A0, A3, F1, F2, F3
(2) SIGNAL; INFORMATION
(3) AMPLITUDE MODULATION
(4) DOUBLE SIDEBAND
(5) SINGLE SIDEBAND
(6) FREQUENCY MODULATION
(7) PHASE MODULATION
(8) CARRIER
(9) SIDEBANDS
(10) BANDWIDTH
(11) ENVELOPE
(12) DEVIATION
(13) OVERMODULATION
(14) SPLATTER
(15) FREQUENCY TRANSLATION; MIXING, MULTIPLICATION
(16) RADIOTELEPRINTING; AUDIO FREQUENCY SHIFT KEYING, MARK, SPACE, SHIFT

I. ANTENNAS AND FEEDLINES

POPULAR AMATEUR RADIO ANTENNAS AND THEIR CHARACTERISTICS:
(1) YAGI ANTENNA
(2) QUAD ANTENNA
(3) PHYSICAL DIMENSIONS
(4) VERTICAL AND HORIZONTAL POLARIZATION
(5) FEEDPOINT IMPEDANCE OF HALF-WAVE DIPOLE QUARTER WAVE VERTICAL
(6) RADIATION PATTERNS; DIRECTIVITY, MAJOR LOBES

CHARACTERISTICS OF POPULAR AMATEUR RADIO ANTENNA FEED LINES; RELATED CONCEPTS:
(7) CHARACTERISTIC IMPEDANCE
(8) STANDING WAVES
(9) STANDING WAVE RATIO; SIGNIFICANCE OF
(10) BALANCED, UNBALANCED
(11) ATTENUATION
(12) ANTENNA-FEEDLINE MISMATCH

Appendix C
FCC Rules and Regulations - Part 97, The Amateur Radio Service

Subpart A—General

Subpart B—Amateur Operator and Station Licenses

OPERATOR LICENSES

OPERATOR LICENSE EXAMINATIONS

STATION LICENSES

CALL SIGNS

DUPLICATE LICENSES AND LICENSE TERM

Subpart C—Technical Standards

Subpart D—Operating Requirements and Procedures

GENERAL

97.78 Practice to be observed by all licensees.
97.79 Control operator requirements.
97.81 Authorized apparatus.
97.82 Availability of operator license.
97.83 Availability of station license.
97.84 Station identification.
97.85 Repeater operation.
97.86 Auxiliary operation.
97.88 Operation of a remotely controlled station.
97.89 Points of communications.
97.91 One-way communications.
97.93 Modulation of carrier.

STATION OPERATION AWAY FROM AUTHORIZED LOCATION

97.95 Operation away from the authorized fixed station location.

SPECIAL PROVISIONS

97.99 Stations used only for radio control of remote model crafts and vehicles.
97.101 Mobile stations aboard ships or aircraft.

LOGS

97.103 Station log requirements.
97.105 Retention of logs.

EMERGENCY OPERATIONS

97.107 Operation in emergencies.

Subpart E—Prohibited Practices and Administrative Sanctions

PROHIBITED TRANSMISSIONS AND PRACTICES

97.112 No remuneration for use of station.
97.113 Broadcasting prohibited.
97.114 Third party traffic.
97.115 Music prohibited.
97.116 Amateur radiocommunication for unlawful purposes prohibited.
97.117 Codes and ciphers prohibited.
97.119 Obscenity, indecency, profanity.
97.121 False signals.
97.123 Unidentified communications.
97.125 Interference.
97.126 Retransmitting radio signals.
97.127 Damage to apparatus.
97.129 Fraudulent licenses.

ADMINISTRATIVE SANCTIONS

97.131 Restricted operation.
97.133 Second notice of same violation.
97.135 Third notice of same violation.
97.137 Answers to notices of violations.

Subpart F—Radio Amateur Civil Emergency Service (RACES)

GENERAL

97.161 Basis and purpose.
97.163 Definitions.
97.165 Applicability of rules.

STATION AUTHORIZATIONS

97.169 Station license required.
97.171 Eligibility for RACES station license.
97.173 Application for RACES station license.
97.175 Amateur radio station registration in civil defense organization.

OPERATING REQUIREMENTS

97.177 Operator requirements.
97.179 Operator privileges.
97.181 Availability of RACES station license and operator licenses.

TECHNICAL REQUIREMENTS

97.185 Frequencies available.
97.189 Points of communications.
97.191 Permissible communications.
97.193 Limitations on the use of RACES stations.

Subpart G—Operation of Amateur Radio Stations in the United States by Aliens Pursuant to Reciprocal Agreements

97.301 Basis, purpose, and scope.
97.303 Permit required.
97.305 Application for permit.
97.307 Issuance of permit.
97.309 Modification, suspension, or cancellation of permit.
97.311 Operating conditions.
97.313 Station identification.

Subpart H—(Reserved)

APPENDICES

1 Examination points.
2 Extracts from Radio Regulations Annexed to the International Telecommunication Convention (Geneva, 1959).
3 Classification of emissions.
4 Convention between the United States of America and Canada, Relating to the Operation by Citizens of Either Country of Certain Radio Equipment or Stations in the Other Country (Effective May 15, 1952).
5 Determination of Antenna Height above Average Terrain.

AUTHORITY: §§ 97.1 to 97.313 issued under 48 Stat. 1066, 1082, as amended; 47 U.S.C. 154, 303. Interpret or apply 48 Stat. 1064-1068, 1081-1105, as amended; 47 U.S.C. Sub-chap. I, III-VI.

SUBPART A—GENERAL

§ 97.1 Basis and purpose.

The rules and regulations in this part are designed to provide an amateur radio service having a fundamental purpose as expressed in the following principles:

(a) Recognition and enhancement of the value of the amateur service to the public as a voluntary noncommercial communication service, particularly with respect to providing emergency communications.

(b) Continuation and extension of the amateur's proven ability to contribute to the advancement of the radio art.

(c) Encouragement and improvement of the amateur radio service through rules which provide for advancing skills in both the communication and technical phases of the art.

(d) Expansion of the existing reservoir within the amateur radio service of trained operators, technicians, and electronics experts.

(e) Continuation and extension of the amateur's unique ability to enhance international good will.

§ 97.3 Definitions.

(a) *Amateur radio service.* A radio communication service of self-training, intercommunication, and technical investigation carried on by amateur radio operators.

(b) *Amateur radio communication.* Noncommercial radio communication by or among amateur radio stations solely with a personal aim and without pecuniary or business interest.

(c) *Amateur radio operator* means a person holding a valid license to operate an amateur radio station issued by the Federal Communications Commission.

(d) *Amateur radio license.* The instrument of authorization issued by the Federal Communications Commission comprised of a station license, and in the case of the primary station, also incorporating an operator license.

Operator license. The instrument of authorization including the class of operator privileges.

Interim Amateur Permit. A temporary operator and station authorization issued to licensees successfully completing Commission supervised examinations for higher class operator licenses.

Station license. The instrument of authorization for a radio station in the Amateur Radio Service.

(e) *Amateur radio station.* A station licensed in the amateur radio service embracing necessary apparatus at a particular location used for amateur radio communication.

(f) *Primary station.* The principal amateur radio station at a specific land location shown on the station license.

(g) *Military recreation station.* An amateur radio station licensed to the person in charge of a station at a land location provided for the recreational use of amateur radio operators, under military auspices of the Armed Forces of the United States.

(h) *Club station.* A separate Amateur radio station licensed to an Amateur radio operator acting as a station trustec for a *bona fide* amateur radio organization or society. A *bona fide* Amateur radio organization or society shall be composed of at least two persons, one of whom must be a licensed Amateur operator, and shall have:

(1) A name,

(2) An instrument of organization (e.g., constitution),

(3) Management, and

(4) A primary purpose which is devoted to Amateur radio activities consistent with § 97.1 and constituting the major portion of the club's activities.

(i) *Space radio station.* An amateur radio station located on an object which is beyond, is intended to go beyond, or has been beyond the major portion of the earth's atmosphere. (Regulations governing this type of station have not yet been adopted and all applications will be considered on an individual basis.)

(j) *Terrestrial location.* Any point within the major portion of the earth's atmosphere, including aeronautical, land, and maritime locations.

(k) *Space location.* (Reserved)

(l) *Amateur radio operation.* Amateur radio communication conducted by amateur radio operators from amateur radio stations, including the following:

Fixed operation. Radio communication conducted from the specific geographical land location shown on the station license.

Portable operation. Radio communication conducted from a specific geographical location other than that shown on the station license.

Mobile operation. Radio communication conducted while in motion or during halts at unspecified locations.

Repeater operation. Radiocommunication, other than auxiliary operation, for retransmitting automatically the radio signals of other amateur radio stations.

Auxiliary operation. Radiocommunication for remotely controlling other amateur radio stations, for automatically relaying the radio signals of other amateur radio stations in a system of stations, or for intercommunicating with other amateur radio stations in a system of amateur radio stations.

(m) *Control* means techniques used for accomplishing the immediate operation of an amateur radio station. Control includes one or more of the following:

(1) *Local control.* Manual control, with the control operator monitoring the operation on duty at the control point located at a station transmitter with the associated operating adjustments directly accessible. (Direct mechanical control, or direct wire control of a transmitter from a control point located on board any aircraft, vessel, or on the same premises on which the transmitter is located, is also considered local control.)

(2) *Remote control.* Manual control, with the control operator monitoring the operation on duty at a control point located elsewhere than at the station transmitter, such that the associated operating adjustments are accessible through a control link.

(3) *Automatic control* means the use of devices and procedures for control so that a control operator does not have to be present at the control point at all times. (Only rules for automatic control of stations in repeater operation have been adopted.)

(n) *Control link.* Apparatus for effecting remote control between a control point and a remotely controlled station.

(o) *Control operator.* An amateur radio operator designated by the licensee of an amateur radio station to also be responsible for the emissions from that station.

(p) *Control point.* The operating position of an amateur radio station where the control operator function is performed.

(q) *Antenna structures.* Antenna structures include the radiating system, its supporting structures, and any appurtenances mounted thereon.

(r) *Antenna height above average terrain.* The height of the center of radiation of an antenna above an averaged value of the elevation above sea level for the surrounding terrain.

(s) *Transmitter.* Apparatus for converting electrical energy received from a source into radio-frequency electromagnetic energy capable of being radiated.

(t) *Effective radiated power.* The product of the radio-frequency power, expressed in watts, delivered to an antenna, and the relative gain of the antenna over that of a half-wave dipole antenna.

(u) *System network diagram.* A diagram showing each station and its relationship to the other stations in a network of stations, and to the control point(s).

(v) *Third-party traffic.* Amateur radio communication by or under the supervision of the control operator at an amateur radio station to another amateur radio station on behalf of anyone other than the control operator.

(w) *Emergency communication.* Any amateur radio communication directly relating to the immediate safety of life of individuals or the immediate protection of property.

(x) *Automatic retransmission.* Retransmission of signals by an amateur radio station whereby the retransmitting station is actuated solely by the presence of a received signal through electrical or electro-mechanical means, i.e., without any direct, positive action by the control operator.

(y) *External radio frequency power amplifier.* Any device which, (1) when used in conjunction with a radio transmitter as a signal source, is capable of amplification of that signal, and (2) is not an integral part of the transmitter as manufactured.

(z) *External radio frequency power amplifier kit.* Any number of electronic parts, usually provided with a schematic diagram or printed circuit board, which, when assembled in accordance with instructions, results in an external radio frequency power amplifier, even if additional parts of any type are required to complete assembly.

SUBPART B—AMATEUR OPERATOR AND STATION LICENSES

OPERATOR LICENSES

§ 97.5 Classes of operator licenses.

Amateur extra class.
Advanced class (previously class A).
General class (previously class B).
Conditional class (previously class C).
Technician class.
Novice class.

§ 97.7 Privileges of operator licenses.

(a) *Amateur Extra Class and Advanced Class.* All authorized amateur privileges including exclusive frequency operating authority in accordance with the following table:

Frequencies	*Class of license authorized*
3500-3525 kHz	Amateur Extra Only
3775-3800 kHz	
7000-7025 kHz	
14,000-14,025 kHz	
21,000-21,025 kHz	
21,250-21,270 kHz	
3800-3890 kHz	Amateur Extra and Advanced.
7150-7225 kHz	
14,200-14,275 kHz	
21,270-21,350 kHz	

(b) *General Class*. All authorized amateur privileges except those exclusive operating privileges which are reserved to the Advanced Class and/or Amateur Extra Class.

(c) *Conditional Class*. Same privileges as General Class. New Conditional Class licenses will not be issued. Present Conditional Class licensees will be issued General Class licenses at time of renewal or modification.

(d) *Technician Class*. All authorized amateur privileges on the frequencies 50.0 MHz and above. Technician Class licenses also convey the full privileges of Novice Class licenses.

(e) *Novice Class*. Radiotelegraphy in the frequency bands 3700-3750 kHz, 7100-7150 kHz (7050-7075 kHz when the terrestrial station location is not within Region 2), 21,100-21,200 kHz, and 28,100-28,200 kHz, using only Type A1 emission.

§ 97.9 Eligibility for new operator license.

Anyone except a representative of a foreign government is eligible for an amateur operator license.

§ 97.11 Application for operator license.

(a) An application (FCC Form 610) for a new operator license, including an application for change in operating privileges, which will require an examination supervised by Commission personnel at a regular Commission examining office shall be submitted to such office in advance of or at the time of the examination, except that, whenever an examination is to be taken at a designated examination point away from a Commission office, the application, together with the necessary filing fee should be submitted in advance of the examination date to the office which has jurisdiction over the examination point involved.

(b) An application (FCC Form 610) for a new operator license, including an application for change in operating privileges, which requests an examination supervised by a volunteer examiner under the provisions of § 97.27, shall be submitted to the FCC field office nearest the applicant. Applications for the Novice Class license should be sent to the Commission's offices in Gettysburg, Pa. 17325. All applications should be accompanied by any necessary filing fee.

(c) An application (FCC Form 610) for renewal and/or modification of license when no change in operating privileges is involved shall be submitted, together with any necessary filing fee, to the Commission's office at Gettysburg, Pennsylvania, 17325.

§ 97.13 Renewal or modification of operator license.

(a) An Amateur operator license may be renewed upon proper application.

(b) The applicant shall qualify for a new license by examination if the requirements of this section are not fulfilled.

(c) Application for renewal and/or modification of an amateur operator license shall be submitted on FCC Form 610 and shall be accompanied by the applicant's license. Application for renewal of unexpired licenses must be made during the license term and should be filed within 90 days but not later than 30 days prior to the end of the license term. In any case in which the licensee has, in accordance with the provisions of this chapter, made timely and sufficient application for renewal of an unexpired license, no license with reference to any activity of a continuing nature shall expire until such application shall have been finally determined.

(d) If a license is allowed to expire, application for renewal may be made during a period of grace of one year after the expiration date. During this one year period of grace, an expired license is not valid. A license renewed during the grace period will be dated currently and will not be backdated to the date of its expiration. Application for renewal shall be submitted on FCC Form 610 and shall be accompanied by the applicant's expired license.

(e) When the name of a licensee is changed or when the mailing address is changed a formal application for modification of license is not required. However, the licensee shall notify the Commission promptly of these changes. The notice, which may be in letter form, shall contain the name and address of the licensee as they appear in the Commission's records, the new name and/or address, as the case may be, the radio station call sign and class of operator license. The notice shall be sent to Federal Communications Commission, Gettysburg, Pa. 17325 and a copy shall be kept by the licensee until a new license is issued.

OPERATOR LICENSE EXAMINATIONS

§ 97.19 When examination is required.

Examination is required for the issuance of a new amateur operator license, and for a change in class of operating privileges. Credit may be given, however, for certain elements of examination as provided in § 97.25.

§ 97.21 Examination elements.

Examinations for amateur operator privileges will comprise one or more of the following examination elements:

(a) Element 1(A): Beginner's code test at five (5) words per minute;

(b) Element 1(B): General code test at thirteen (13) words per minute;

(c) Element 1(C): Expert's code test at twenty (20) words per minute;

(d) Element 2: Basic law comprising rules and regulations essential to beginners' operation, including sufficient elementary radio theory for the understanding of those rules;

(e) Element 3: General amateur practice and regulations involving radio operation and apparatus and provisions of treaties, statutes, and rules affecting amateur stations and operators;

(f) Element 4(A): Intermediate amateur practice involving intermediate level radio theory and operation as applicable to modern amateur techniques, including, but not limited to, radiotelephony and radiotelegraphy;

(g) Element 4(B): Advanced amateur practice involving advanced radio theory and operation as applicable to modern amateur techniques, including, but not limited to, radiotelephony, radiotelegraphy, and transmissions of energy for measurements and observations applied to propagation, for the radio control of remote objects and for similar experimental purposes.

§ 97.23 Examination requirements.

Applicants for operator licenses will be required to pass the following examination elements:

(a) Amateur Extra Class: Elements 1(C), 2, 3, 4(A) and 4(B);

(b) Advanced Class: Elements 1(B), 2, 3, and 4(A);

(c) General Class: Elements 1(B), 2 and 3;

(d) Technician Class: Elements 1(A), 2, and 3;

(e) Novice Class: Elements 1(A) and 2.

§ 97.25 Examination credit.

(a) An applicant for a higher class of amateur operator license who holds any valid amateur license will be required to pass only those elements of the higher class examination that are not included in the examination for the amateur license held.

(b) Upon presentation of a properly completed Amateur Code Credit Certificate, FCC Form 845, the FCC shall give the applicant for an amateur radio operator license examination credit for the code speed listed on the Amateur Code Credit Certificate. An Amateur Code Credit Certificate is valid for a period of one year from the date of its issuance and will be honored only at the FCC field office that issued the Amateur Code Credit Certificate.

(c) An applicant for an amateur operator license will be given credit for either telegraph code element 1(A) or 1(B) if within 5 years prior to the receipt of his application by the Commission he held a commercial radiotelegraph operator license or permit issued by the Federal Communications Commission. An applicant for an amateur extra class license will be given credit for the telegraph code element 1(C) if he holds a valid first class commercial radiotelegraph operator license or permit issued by the Federal Communications Commission or holds any commercial radiotelegraph operator license or permit issued by the Federal Communications Commission containing an aircraft radiotelegraph endorsement.

(d) An applicant for the amateur extra class operator license will be given credit for examination element 1(C) if he so requests and submits evidence of having held the amateur extra first class license, having continuously held its successor license. An applicant should present his proof in advance of the desired examination time to the Chief, Personal Radio Division, Washington, D.C. 20554 and receive a letter of certification for presentation to the field office where the examination will be taken. No code credit will be given without the letter of certification.

(e) No examination credit, except as herein provided, shall be allowed on the basis of holding or having held any amateur or commercial operator license.

§ 97.27 Mail examinations for applicants unable to travel.

The Commission may permit the examinations for an Amateur Extra, Advanced, General, or Technician Class license to be administered at a location other than a Commission examination point by an examiner chosen by the Commission when it is shown by physician's certification that the applicant is unable to appear at a regular Commission examination point because of a protracted disability preventing travel.

§ 97.28 Manner of conducting examinations.

(a) Except as provided in §97.27, all examinations for Amateur Extra, Advanced,General, and Technician Class operator licenses will be conducted by authorized Commission personnel or representatives at locations and times specified by the Commission. Examination elements given under the provisions of §97.27 will be administered by an examiner selected by the Commission. All applications for consideration of eligibility under §97.27 should be filed on FCC Form 610, and should be sent to the FCC field office nearest the applicant. (A list of these offices appears in §0.121 of the Commission's Rules and can be obtained from the Regional Services Division, Field Operations Bureau, FCC, Washington, D.C. 20554, or any field office.)

(b) The examination for a Novice Class operator license shall be conducted and supervised by a volunteer examiner selected by the applicant, unless otherwise prescribed by the Commission. The volunteer examiner shall be at least 18 years of age, shall be unrelated to the applicant, and shall be the holder of an Amateur Extra, Advanced, or General Class operator license. The written portion of the Novice Class operator examination shall be obtained, administered, and submitted in accordance with the following procedure:

(1) Within 10 days after successfully completing telegraphy examination element 1(A), an applicant shall submit an application (FCC Form 610) to the Commission's office in Gettysburg, Pennsylvania 17325. The application shall include a written request from the volunteer examiner for the examination papers for Element 2. The examiner's written request shall include (i) the names and permanent addresses of the examiner and the applicant, (ii) a description of the examiner's qualifications to administer the examination, (iii) the examiner's statement that the applicant has passed telegraphy element 1(A) under his supervision within the 10 days prior to submission of the request, and (iv) the examiner's written signature. Examination papers will be forwarded only to the volunteer examiner.

(2) The volunteer examiner shall be responsible for the proper conduct and necessary supervision of the examination. Administration of the examination shall be in accordance with the instructions included with the examination papers.

(3) The examination papers, either completed or unopened in the event the examination is not taken, shall be returned by the volunteer examiner to the Commission's office in Gettysburg, Pa., no later than 30 days after the date the papers are mailed by the Commission (the date of mailing is normally stamped by the Commission on the outside of the examination envelope).

(c) The code test required of an applicant for an amateur radio operator license, in accordance with the provisions of §§97.21 and 97.23 shall determine the applicant's ability to transmit by hand key (straight key or, if supplied by the applicant, any other type of hand operated key such as a semi-automatic or electronic key, but not a keyboard keyer) and to receive by ear, in plain language, messages in the international Morse code at not less than the prescribed speed during a five minute test period. Each five characters shall be counted as one word. Each punctuation mark and numeral shall be counted as two characters.

(d) All written portions of the examinations for amateur operator privileges shall be completed by the applicant in legible handwriting or hand printing. Whenever the applicant's signature is required, his normal signature shall be used. Applicants unable to comply with these requirements, because of physical disability, may dictate their answers to the examination questions and the receiving code test. If the examination or any part thereof is dictated, the examiner shall certify the nature of the applicant's disability and the name and address of the person(s) taking and transcribing the applicant's dictation.

§ 97.31 Grading of examinations.

(a) Code tests for sending and receiving are graded separately.

(b) Seventy-four percent (74%) is the passing grade for written examinations. For the purpose of grading, each element required in qualifying for a particular license will be considered as a separate examination. All written examinations will be graded only by Commission personnel.

§ 97.32 Interim Amateur Permits.

(a) Upon successful completion of a Commission supervised Amateur Radio Service operator examination, an applicant already licensed in the Amateur Radio Service may operate his amateur radio station pending issuance of his permanent amateur operator and station licenses under the terms and conditions of an Interim Amateur Permit, evidenced by a properly executed FCC Form 660-B.

(b) An Interim Amateur Permit conveys all operating privileges of the applicant's new operator license classification.

(c) The transmissions of amateur radio stations operated under the authority of Interim Amateur Permits shall be identified in the manner specified in §97.84.

(d) The original Interim Amateur Permit of an amateur radio operator shall be kept in the personal possession of or posted in a conspicuous place in the room occupied by such operator when operating an amateur radio station under the authority of an Interim Amateur Permit.

(e) Interim Amateur Permits are valid for a period of 90 days from the date of issuance or until issuance of the permanent station and operator licenses, whichever comes first, but may be set aside by the Commission within the 90 day term if it appears that the permanent operator and station licenses cannot be granted routinely.

(f) Interim Amateur Permits shall not be renewed.

§ 97.33 Eligibility for re-examination.

An applicant who fails an examination element required for an amateur radio operator license shall not apply to be examined for the same or higher examination element within thirty days of the date the examination element was failed.

Station Licenses

§ 97.37 General eligibility for station license.

An amateur radio station license will be issued only to a licensed amateur radio operator, except that a military recreation station license may also be issued to an individual not licensed as an amateur radio operator (other than a representative of a foreign government), who is in charge of a proposed military recreation station not operated by the U.S. Government but which is to be located in approved public quarters.

§ 97.39 Eligibility of corporations or organizations to hold station license.

An amateur station license will not be issued to a school, company, corporation, association, or other organization, except that in the case of a *bona fide* amateur radio organization or society meeting the criteria set forth in Section 97.3, a station license may be issued to a licensed amateur operator, other than the holder of a Novice Class license, as trustee for such society.

§ 97.40 Station license required.

(a) No transmitting station shall be operated in the amateur radio service without being licensed by the Federal Communications Commission.

(b) Every amateur radio operator shall have one, but only one, primary amateur radio station license.

§ 97.41 Application for station license.

(a) Each application for a club or military recreation station license in the Amateur Radio Service shall be made on the FCC Form 610-B. Each application for any other amateur radio license shall be made on the FCC Form 610.

(b) One application and all papers incorporated therein and made a part thereof shall be submitted for each amateur station license. If the application is only for a station license, it shall be filed directly with the Commission's Gettysburg, Pennsylvania office. If the application also contains an application for any class of amateur operator license, it shall be filed in accordance with the provisions of §97.11.

(c) Each applicant in the Safety and Special Radio Services (1) for modification of a station license involving a site change or a substantial increase in tower height or (2) for a license for a new station must, before commencing construction, supply the environmental information, where required, and must follow the procedure prescribed by Subpart I of Part 1 of this chapter (§§ 1.1301 through 1.1319) unless Commission action authorizing such construction would be a minor action with the meaning of Subpart I of Part I.

§ 97.42 Mailing address furnished by licensee.

Except for applications submitted by Canadian citizens pursuant to agreement between the United States and Canada (TIAS No. 2508 and No. 6931), each application shall set forth and each licensee shall furnish the Commission with an address in the United States to be used by the Commission in serving documents or directing correspondence to that licensee. Unless any licensee advises the Commission to the contrary, the address contained in the licensee's most recent application will be used by the Commission for this purpose.

§ 97.43 Location of station.

Every amateur radio station shall have one land location, the address of which appears on the station license, and at least one control point.

§ 97.45 Limitations on antenna structures.

(a) Except as provided in paragraph (b) of this section, an antenna for a station in the Amateur Radio Service which exceeds the following height limitations may not be erected or used unless notice has been filed with both the FAA on FAA Form 7460-1 and with the Commission on Form 714 or on the license application form, and prior approval by the Commission has been obtained for:

(1) Any construction or alteration of more than 200 feet in height above ground level at its site (§17.7(a) of this chapter).

(2) Any construction or alteration of greater height than an imaginary surface extending outward and upward at one of the following slopes (§17.7(b) of this chapter):

(i) 100 to 1 for a horizontal distance of 20,000 feet from the nearest point of the nearest runway of each airport with at least one runway more than 3,200 feet in length, excluding heliports and seaplane bases without specified boundaries, if that airport is either listed in the Airport Directory of the current Airman's Information Manual or is operated by a Federal military agency.

(ii) 50 to 1 for a horizontal distance of 10,000 feet from the nearest point of the nearest runway of each airport with its longest runway no more than 3,200 feet in length, excluding heliports and seaplane bases without specified boundaries, if that airport is either listed in the Airport Directory or is operated by a Federal military agency.

(iii) 25 to 1 for a horizontal distance of 5,000 feet from the nearest point of the nearest landing and takeoff area of each heliport listed in the Airport Directory or operated by a Federal military agency.

(3) Any construction or alteration on an airport listed in the Airport Directory of the Airman's Information Manual (§17.7(c) of this chapter).

(b) A notification to the Federal Aviation Administration is not required for any of the following construction or alteration:

(1) Any object that would be shielded by existing structures of a permanent and substantial character or by natural terrain or topographic features of equal or greater height, and would be located in the congested area of a city, town, or settlement where it is evident beyond all reasonable doubt that the structure so shielded will not adversely affect safety in air navigation. Applicants claiming such exemption shall submit a statement with their application to the Commission explaining the basis in detail for their finding (§17.14(a) of this chapter).

(2) Any antenna structure of 20 feet or less in height except one that would increase the height of another antenna structure (§17.14(b) of this chapter).

(c) Further details as to whether an aeronautical study and/or obstruction marking and lighting may be required, and specifications for obstruction marking and lighting when required, may be obtained from Part 17 of this chapter, "Construction, Marking, and Lighting of Antenna Structures." Information regarding the inspection and maintenance of antenna structures requiring obstruction marking and lighting is also contained in Part 17 of this chapter.

§ 97.47 Renewal and/or modification of amateur station license.

(a) Application for renewal and/or modification of an individual station license shall be submitted on FCC Form 610, and application for renewal and/or modification of an amateur club or military recreation station shall be submitted on FCC Form 610-B. In every case the application shall be accompanied by the applicant's license or photocopy thereof. Applications for renewal of unexpired licenses must be made during the license term and should be filed not later than 60 days prior to the end of the license term. In any case in which the licensee has in accordance with the provisions of this chapter, made timely and sufficient application for renewal of an unexpired license, no license with reference to any activity of a continuing nature shall expire until such application shall have been finally determined.

(b) If a license is allowed to expire, application for renewal may be made during a period of grace of 1 year after the expiration date. During this 1-year period of grace, an expired license is not valid. A license renewed during the grace period will be dated currently and will not be backdated to the date of expiration. An application for an individual station license shall be submitted on FCC Form 610. An application for an amateur club or military recreation station license shall be submitted on FCC Form 610-B. In every case the application shall be accompanied by the applicant's expired license or a photocopy thereof.

(c) When the name of a licensee is changed (without changes in the ownership, control, or corporate structure), or when the mailing address is changed (without changing the authorized location of the amateur radio station) a formal application for modification of license is not required. However, the licensee shall notify the Commission promptly of these changes. The notice, which may be in letter form, shall contain the name and address of the licensee as they appear in the Commission's records, the new name and/or address, as the case may be, and the call sign and the class of operator license. The notice shall be sent to Federal Communications Commission, Gettysburg, Pa., 17325, and a copy shall be maintained with the license of each station until a new license is issued.

§ 97.49 Commission modification of station license.

(a) Whenever the Commission shall determine that public interest, convenience, and necessity would be served, or any treaty ratified by the United States will be more fully complied with, by the modification of any radio station license either for a limited time, or for the duration of the term thereof, it shall issue an order for such licensee to show cause why such license should not be modified.

(b) Such order to show cause shall contain a statement of the grounds and reasons for such proposed modification, and shall specify wherein the said license is required to be modified. It shall require the licensee against whom it is directed to appear at a place and time therein named, in no event to be less than 30 days from the date of receipt of the order, to show cause why the proposed modification should not be made and the order of modification issued.

(c) If the licensee against whom the order to show cause is directed does not appear at the time and place provided in said order, a final order of modification shall issue forthwith.

CALL SIGNS

§ 97.51 Assignment of call signs.

(a) The Commission shall assign the call sign of an amateur radio station on a systematic basis.

(b) The Commission shall not grant any request for a specific call sign.

(c) From time to time the Commission will issue public announcements detailing the policies and procedures governing the systematic assignment of call signs and any changes in those policies and procedures.

DUPLICATE LICENSES AND LICENSE TERM

§ 97.57 Duplicate license.

Any licensee requesting a duplicate license to replace an original which has been lost, mutilated, or destroyed, shall submit a statement setting forth the facts regarding the manner in which the original license was lost, mutilated, or destroyed. If, subsequent to receipt by the licensee of the duplicate license, the original license is found, either the duplicate or the original license shall be returned immediately to the Commission.

§ 97.59 License term.

(b) Amateur station licenses are normally valid for a period of five years from the date of issuance of a new or renewed license. All amateur station licenses, regardless of when issued, will expire on the same date as the licensee's amateur operator license.

(c) A duplicate license or a modified license which is not being renewed shall bear the same expiration date as the license for which it is a modification or duplicate.

SUBPART C—TECHNICAL STANDARDS

§ 97.61 Authorized frequencies and emissions.

(a) The following frequency bands and associated emissions are available to amateur radio stations for amateur radio operation, other than repeater operation and auxiliary operation, subject to the limitations of §97.65 and paragraph (b) of this section:

Frequency band	Emissions	Limitations (See paragraph (b))
kHz		
1800-2000	A1, A3	1,2
3500-4000	A1	
3500-3775	F1	
3775-3890	A5, F5	
3775-4000	A3, F3	4
4383.8	A3J/A3A	13
7000-7300	A1	3,4
7000-7150	F1	3,4
7075-7100	A3, F3	11
7150-7225	A5, F5	3,4
7150-7300	A3, F3	3,4
14000-14350	A1	
14000-14200	F1	
14200-14275	A5, F5	
14200-14350	A3, F3	
MHz		
21.000-21.450	A1	
21.000-21.250	F1	
21.250-21.350	A5, F5	
21.250-21.450	A3, F3	
28.000-29.700	A1	
28.000-28.500	F1	
28.500-29.700	A3, F3, A5, F5	
50.0-54.0	A1	
50.1-54.0	A2, A3, A4, A5, F1, F2, F3, F5	
51.0-54.0	A0	
144-148	A1	
144.1-148.0	A0, A2, A3, A4, A5, F0, F1, F2, F3, F5	
220-225	A0, A1, A2, A3, A4, A5, F0, F1, F2, F3, F4, F5	
420-450	A0, A1, A2, A3, A4, A5, F0, F1, F2, F3, F4, F5	5,7
1215-1300	A0, A1, A2, A3, A4, A5, F0, F1, F2, F3, F4, F5	5
2300-2450	A0, A1, A2, A3, A4, A5, F0, F1, F2, F3, F4, F5, P	5,8
3300-3500	A0, A1, A2, A3, A4, A5, F0, F1, F2, F3, F4, F5, P	5,12
5650-5925	A0, A1, A2, A3, A4, A5, F0, F1, F2, F3, F4, F5, P	5,9
GHz		
10.000-10.500	A0, A1, A2, A3, A4, A5, F0, F1, F2, F3, F4, F5	5
24.000-24.250	A0, A1, A2, A3, A4, A5, F0, F1, F2, F3, F4, F5, P	5,10
48.000-50.000	A0, A1, A2, A3, A4, A5, F0, F1, F2, F3, F4, F5, P	
71.000-76.000	A0, A1, A2, A3, A4, A5, F0, F1, F2, F3, F4, F5, P	
165.000-170.000	A0, A1, A2, A3, A4, A5, F0, F1, F2, F3, F4, F5, P	
240.000-250.000	A0, A1, A2, A3, A4, A5, F0, F1, F2, F3, F4, F5, P	
Above 300.000	A0, A1, A2, A3, A4, A5, F0, F1, F2, F3, F4, F5, P	

(b) Limitations:

(1) The use of frequencies in this band is on a shared basis with the LORAN-A radionavigation system and is subject to cancellation or revision, in whole or in part, by order of the Commission, without hearing, whenever the Commission shall determine such action is necessary in view of the priority of the LORAN-A radionavigation system. The use of these frequencies by amateur stations shall not cause harmful interference to LORAN-A system. If an amateur station causes such interference, operation on the frequencies involved must cease if so directed by the Commission.

(2) Operation shall be limited to:

Area	Maximum DC plate input power in watts							
	1800-1825 kHz	1825-1850 kHz	1850-1875 kHz	1875-1900 kHz	1900-1925 kHz	1925-1950 kHz	1950-1975 kHz	1975-2000 kHz
	Day/Night	Day/Night	Day/Night	Day/Night	Day/Night	Day/Night	Day/Night	Day/Night
Alabama	500/100	100/25	0	0	0	0	100/25	500/100
Alaska	1000/200	500/100	500/100	100/25	0	0	0	0
Arizona	1000/200	500/100	500/100	0	0	0	0	0
Arkansas	1000/200	500/100	100/25	0	0	100/25	100/25	500/100
California	1000/200	500/100	500/100	100/25	0	0	0	0
Colorado	1000/200	500/100	200/50	0	0	0	0	200/50
Connecticut	500/100	100/25	0	0	0	0	0	0
Delaware	500/100	100/25	0	0	0	0	0	100/25
District of Columbia	500/100	100/25	0	0	0	0	0	100/25
Florida	500/100	100/25	0	0	0	0	100/25	500/100
Georgia	500/100	100/25	0	0	0	0	0	200/50
Hawaii	0	0	0	0	200/50	100/25	100/25	500/100
Idaho	1000/200	500/100	500/100	100/25	100/25	100/25	100/25	500/100
Illinois	1000/200	500/100	100/25	0	0	0	0	200/50
Indiana	1000/200	500/100	100/25	0	0	0	0	200/50
Iowa	1000/200	500/100	200/50	0	0	100/25	100/25	500/100
Kansas	1000/200	500/100	100/25	0	0	100/25	100/25	500/100
Kentucky	1000/200	500/100	100/25	0	0	0	0	200/50
Louisiana	500/100	100/25	0	0	0	0	100/25	500/100
Maine	500/100	100/25	0	0	0	0	0	0
Maryland	500/100	100/25	0	0	0	0	0	100/25
Massachusetts	500/100	100/25	0	0	0	0	0	0
Michigan	1000/200	500/100	100/25	0	0	0	0	100/25
Minnesota	1000/200	500/100	500/100	100/25	100/25	100/25	100/25	500/100
Mississippi	500/100	100/25	0	0	0	0	100/25	500/100
Missouri	1000/200	500/100	100/25	0	0	100/25	100/25	500/100
Montana	1000/200	500/100	500/100	100/25	100/25	100/25	100/25	500/100
Nebraska	1000/200	500/100	200/50	0	0	100/25	100/25	500/100
Nevada	1000/200	500/100	500/100	100/25	0	0	0	0
New Hampshire	500/100	100/25	0	0	0	0	0	0
New Jersey	500/100	100/25	0	0	0	0	0	0
New Mexico	1000/200	500/100	100/25	0	0	100/25	500/100	1000/200
New York	500/100	100/25	0	0	0	0	0	0
North Carolina	500/100	100/25	0	0	0	0	0	100/25
North Dakota	1000/200	500/100	500/100	100/25	100/25	100/25	100/25	500/100
Ohio	1000/200	500/100	100/25	0	0	0	0	100/25
Oklahoma	1000/200	500/100	100/25	0	0	100/25	100/25	500/100
Oregon	1000/200	500/100	500/100	100/25	0	0	0	0

Pennsylvania	500/100	100/25	0	0	0	0	0	0
Rhode Island	500/100	100/25	0	0	0	0	0	0
South Carolina	500/100	100/25	0	0	0	0	0	200/50
South Dakota	1000/200	500/100	500/100	100/25	100/25	100/25	100/25	500/100
Tennessee	1000/200	500/100	100/25	0	0	0	0	200/50
Texas	500/100	100/25	0	0	0	0	0	200/50
Utah	1000/200	500/100	500/100	100/25	100/25	0	0	100/25
Vermont	500/100	100/25	0	0	0	0	0	0
Virginia	500/100	100/25	0	0	0	0	0	100/25
Washington	1000/200	500/100	500/100	100/25	0	0	0	0
West Virginia	1000/200	500/100	100/25	0	0	0	0	100/25
Wisconsin	1000/200	500/100	200/50	0	0	0	0	200/50
Wyoming	1000/200	500/100	500/100	100/25	100/25	0	0	200/50
Puerto Rico	500/100	100/25	0	0	0	0	0	200/50
Virgin Islands	500/100	100/25	0	0	0	0	0	200/50
Swan Island	500/100	100/25	0	0	0	0	100/25	500/100
Serrana Bank	500/100	100/25	0	0	0	0	100/25	500/100
Roncador Key	500/100	100/25	0	0	0	0	100/25	500/100
Navassa Island	500/100	100/25	0	0	0	0	0	200/50
Baker, Canton, Enderbury, Howland	100/25	0	0	100/25	100/25	0	0	100/25
Guam, Johnston, Midway	0	0	0	0	100/25	0	0	100/25
American Samoa	200/50	0	0	200/50	200/50	0	0	200/50
Wake	100/25	0	0	100/25	0	0	0	0
Palmyra, Jarvis	0	0	0	0	200/50	0	0	200/50

(3) Where, in adjacent regions or subregions, a band of frequencies is allocated to different services of the same category, the basic principle is the equality of right to operate. Accordingly, the stations of each service in one region or subregion must operate so as not to cause harmful interference to services in the other regions or subregions (No. 117, the Radio Regulations, Geneva, 1959).

(4) 3900-4000 kHz and 7100-7300 kHz are not available in the following U.S. possessions: Baker, Canton, Enderbury, Guam, Howland, Jarvis, Palmyra, American Samoa, and Wake Islands.

(5) Amateur stations shall not cause interference to the Government radiolocation service.

(6) (Reserved)

(7) In the following areas the d.c. plate input power to the final transmitter stage shall not exceed 50 watts, except when authorized by the appropriate Commission Engineer in Charge and the appropriate Military Area Frequency Coordinator.

(i) Those portions of Texas and New Mexico bounded by latitude 33°24′ N., 31°53′ N., and longitude 105°40′ W. and 106°40′ W.

(ii) The State of Florida, including the Key West area and the areas enclosed within circles of 200-mile radius centered at 28°21′ N., 80°43′W. and 30°30′ N., 86°30′ W.

(iii) The State of Arizona.

(iv) Those portions of California and Nevada south of latitude 37°10′ N. and the area within a 200-mile radius of 34°09′ N., 119°11′ W.

(8) No protection in the band 2400-2500 MHz is afforded from interference due to the operation of industrial, scientific, and medical devices on 2450 MHz.

(9) No protection in the band 5725-5875 MHz is afforded from interference due to the operation of industrial, scientific and medical devices on 5800 MHz.

(10) No protection in the band 24.00-24.25 GHz is afforded from interference due to the operation of industrial, scientific and medical devices on 24.125 GHz.

(11) The use of A3 and F3 in this band is limited to amateur radio stations located outside Region 2.

(12) Amateur stations shall not cause interference to the Fixed-Satellite Service operating in the band 3400-3500 MHz.

(13) The frequency 4383.8 kHz, maximum power 150 watts, may be used by any station authorized under this part to communicate with any other station authorized in the State of Alaska for emergency communications. No airborne operations will be permitted on this frequency. Additionally, all stations operating on this frequency must be located in or within 50 nautical miles of the State of Alaska.

(c) All amateur frequency bands above 29.5 MHz are available for repeater operation, except 50.0-52.0 MHz, 144.0-144.5 MHz, 145.5-146.0 MHz, 220.0-220.5 MHz, 431.0-433.0 MHz, and 435.0-438.0 MHz. Both the input (receiving) and output (transmitting) frequencies of a station in repeater operation shall be frequencies available for repeater operation.

(d) All amateur frequency bands above 220.5 MHz, except 431-433 MHz, and 435-438 MHz, are available for auxiliary operation.

§ 97.63 Selection and use of frequencies.

(a) An amateur station may transmit on any frequency within any authorized amateur frequency band.

(b) Sideband frequencies resulting from keying or modulating a carrier wave shall be confined within the authorized amateur band.

(c) The frequencies available for use by a control operator of an amateur station are dependent on the operator license classification of the control operator and are listed in §97.7.

§ 97.65 Emission limitations.

(a) Type A0 emission, where not specifically designated in the bands listed in §97.61, may be used for short periods of time when required for authorized remote control purposes or for experimental purposes. However, these limitations do not apply where type A0 emission is specifically designated.

(b) Whenever code practice, in accordance with §97.91(d), is conducted in bands authorized for A3 emission, tone modulation of the radiotelephone transmitter may be utilized when interspersed with appropriate voice instructions.

(c) On frequencies below 29.0 MHz and between 50.1 and 52.5 MHz, the bandwidth of an F3 emission (frequency or phase modulation) shall not exceed that of an A3 emission having the same audio characteristics; and the purity and stability of emissions shall comply with the requirements of §97.73.

(d) On frequencies below 50 MHz, the bandwidth of A5 and F5 emissions shall not exceed that of an A3 single sideband emission.

(e) On frequencies between 50 MHz and 225 MHz, single sideband or double sideband A5 emission may be used and the bandwidth shall not exceed that of an A3 single sideband or double sideband signal respectively. The bandwidth of F5 emission shall not exceed that of an A3 single sideband emission.

(f) Below 225 MHz, A3 and A5 emissions may be used simultaneously on the same carrier frequency provided the total bandwidth does not exceed that of an A3 double sideband emission.

§ 97.67 Maximum authorized power.

(a) Except for power restrictions as set forth in §97.61 and paragraph (d) below each amateur transmitter may be operated with a power input not exceeding one kilowatt to the plate circuit of the final amplifier stage of an amplifier oscillator transmitter or to the plate circuit of an oscillator transmitter. An amateur transmitter operating with a power input exceeding 900 watts to the plate circuit shall provide means for accurately measuring the plate power input to the vacuum tube or tubes supplying power to the antenna.

(b) Notwithstanding the provisions of paragraph (a) of this section, amateur stations shall use the minimum amount of transmitter power necessary to carry out the desired communications.

(c) Within the limitations of paragraphs (a) and (b) of this section, the effective radiated power of an amateur radio station in repeater operation shall not exceed the power specified for the antenna height above average terrain in the following table:

Antenna height above average terrain	Maximum effective radiated power for frequency bands above:			
	52 MHz	144.5 MHz	420 MHz	1215 MHz
Below 50 feet	100 watts	800 watts	Paragraphs (a) and (b)	Paragraphs (a) and (b)
50-99 feet	100 watts	400 watts	..do..	..do..
100-499 feet	50 watts	400 watts	800 watts	..do..
500-999 feet	25 watts	200 watts	800 watts	..do..
Above 1000 feet	25 watts	100 watts	400 watts	..do..

(d) In the frequency bands 3700-3750 kHz, 7100-7150 kHz (7050-7075 kHz when the terrestrial location of the station is not within Region 2), 21,100-21,200 kHz and 28,100-28,200 kHz, the power input to the transmitter final amplifying stage supplying radio frequency energy to the antenna shall not exceed 250 watts, exclusive of power for heating the cathode of a vacuum tube(s).

§ 97.69 Radio teleprinter transmissions.

The following special conditions shall be observed during the transmission of radio teleprinter signals on authorized frequencies by amateur stations:

(a) A single channel five-unit (start-stop) teleprinter code shall be used which shall correspond to the International Telegraphic Alphabet No. 2 with respect to all letters and numerals (including the slant sign or fraction bar) but special signals may be employed for the remote control of receiving printers, or for other purposes, in "figures" positions not utilized for numerals. In general, this code shall conform as nearly as possible to the teleprinter code or codes in common commercial usage in the United States.

(b) The normal transmitting speed of the radio teleprinter signal keying equipment shall be adjusted as closely as possible to one of the standard teleprinter speeds, namely, 60 (45 bauds), 67 (50 bauds), 75 (56.25 bauds) or 100 (75 bauds) words per minute, and in any event, within the range of ±5 words per minute of the selected standard speed.

(c) When frequency shift keying (type F1 emission) is utilized, the deviation in frequency from the mark signal to space signal, or from the space signal to the mark signal, shall be less than 900 Hertz.

(d) When audio frequency shift keying (type A2 or type F2 emission) is utilized, the highest fundamental modulating audio frequency shall not exceed 3000 hertz, and the difference between the modulating audio frequency for the mark signal and that for the space signal shall be less than 900 hertz.

§ 97.71 Transmitter power supply.

The licensee of an amateur station using frequencies below 144 megahertz shall use adequately filtered direct-current plate power supply for the transmitting equipment to minimize modulation from this source.

§ 97.73 Purity of emissions.

(a) Except for a transmitter or transceiver built before April 15, 1977 or first marketed before January 1, 1978, the mean power of any spurious emission or radiation from an amateur transmitter, transceiver, or external radio frequency power amplifier being operated with a carrier frequency below 30 MHz shall be at least 40 decibels below the mean power of the fundamental without exceeding the power of 50 milliwatts. For equipment of mean power less than five watts, the attenuation shall be at least 30 decibels.

(b) Except for a transmitter or transceiver built before April 15, 1977 or first marketed before January 1, 1978, the mean power of any spurious emission or radiation from an amateur transmitter, transceiver, or external radio frequency power amplifier being operated with a carrier frequency above 30 MHz but below 235 MHz shall be at least 60 decibels below the mean power of the fundamental. For a transmitter having a mean power of 25 watts or less, the mean power of any spurious radiation supplied to the antenna transmission line shall be at least 40 decibels below the mean power of the fundamental without exceeding the power of 25 microwatts, but need not be reduced below the power of 10 microwatts.

(c) Paragraphs (a) and (b) of this section notwithstanding, all spurious emissions or radiation from an amateur transmitter, transceiver, or external radio frequency power amplifier shall be reduced or eliminated in accordance with good engineering practice.

(d) If any spurious radiation, including chassis or power line radiation, causes harmful interference to the reception of another radio station, the licensee may be required to take steps to eliminate the interference in accordance with good engineering practice.

NOTE: For the purposes of this section, a spurious emission or radiation means any emission or radiation from a transmitter, transceiver, or external radio frequency power amplifier which is outside of the authorized Amateur Radio Service frequency band being used.

§ 97.74 Frequency measurement and regular check.

The licensee of an amateur station shall provide for measurement of the emitted carrier frequency or frequencies and shall establish procedures for making such measurement regularly. The measurement of the emitted carrier frequency or frequencies shall be made by means independent of the means used to control the radio frequency or frequencies generated by the transmitting apparatus and shall be of sufficient accuracy to assure operation within the amateur frequency band used.

§ 97.75 Use of external radio frequency (RF) power amplifiers.

(a) Until April 28, 1981, any external radio frequency (RF) power amplifier used or attached at any amateur radio station shall be type accepted in accordance with Subpart J of Part 2 of the FCC's Rules for operation in the Amateur Radio Service, unless one or more of the following conditions are met:

(1) The amplifier is not capable of operation on any frequency or frequencies below 144 MHz (the amplifier shall be considered incapable of operation below 144 MHz if the mean output power decreases, as frequency decreases from 144 MHz, to a point where 0 decibels or less gain is exhibited at 120 MHz and below and the amplifier is not capable of being easily modified to provide amplification below 120 MHz);

(2) The amplifier was originally purchased before April 28, 1978;

(3) The amplifier was—

(i) Constructed by the licensee, not from an external RF power amplifier kit, for use at his amateur radio station;

(ii) Purchased by the licensee as an external RF power amplifier kit before April 28, 1978 for use at his amateur radio station; or

(iii) Modified by the licensee for use at his amateur radio station in accordance with §2.1001 of the FCC's Rules;

(4) The amplifier was purchased by the licensee from another amateur radio operator who—

(i) Constructed the amplifier, but not from an external RF power amplifier kit;

(ii) Purchased the amplifier as an external RF power amplifier kit before April 28, 1978 for use at his amateur radio station; or

(iii) Modified the amplifier for use at his amateur radio station in accordance with §2.1001 of the FCC's Rules;

(5) The external RF power amplifier was purchased from a dealer who obtained it from an amateur radio operator who—

(i) Constructed the amplifier, but not from an external RF power amplifier kit;

(ii) Purchased the amplifier as an external RF power amplifier kit before April 28, 1978 for use at his amateur radio station; or

(iii) Modified the amplifier for use at his amateur radio station in accordance with §2.1001 of the FCC's Rules; or

(6) The amplifier was originally purchased after April 27, 1978 and has been issued a marketing waiver by the FCC.

(b) A list of type accepted equipment may be inspected at FCC headquarters in Washington, D.C. or at any FCC field office. Any external RF power amplifier appearing on this list as type accepted for use in the Amateur Radio Service may be used in the Amateur Radio Service.

NOTE: No more than one unit of one model of an external RF power amplifier shall be constructed or modified during any calendar year by an amateur radio operator for use in the Amateur Radio Service without a grant of type acceptance.

§ 97.76 Requirements for type acceptance of external radio frequency (RF) power amplifiers and external radio frequency power amplifier kits.

(a) Until April 28, 1981, any external radio frequency (RF) power amplifier or external RF power amplifier kit marketed (as defined in §2.815), manufactured, imported or modified for use in the Amateur Radio Service shall be type accepted for use in the Amateur Radio Service in accordance with Subpart J or Part 2 of the FCC's Rules. This requirement does not apply if one or more of the following conditions are met:

(1) The amplifier is not capable of operation on any frequency or frequencies below 144 MHz (the amplifier shall be considered incapable of operation below 144 MHz if the mean output power decreases, as frequency decreases from 144 MHz, to a point where 0 decibels or less gain is exhibited at 120 MHz and below and the amplifier is not capable of being easily modified to provide amplification below 120 MHz).

(2) The amplifier was originally purchased before April 28, 1978 by an amateur radio operator for use at his amateur radio station;

(3) The amplifier was constructed or modified by an amateur radio operator for use at his amateur radio station in accordance with §2.1001 of the FCC's Rules;

(4) The amplifier was constructed or modified by an amateur radio operator in accordance with §2.1001 of the FCC's Rules and sold to another amateur radio operator or to a dealer;

(5) The amplifier was constructed or modified by an amateur radio operator in accordance with §2.1001 of the FCC's Rules and sold by a dealer to an amateur radio operator for use at his amateur radio station; or

(6) The amplifier was manufactured before April 28, 1978 and has been issued a marketing waiver by the FCC.

(b) No more than one unit of one model of an external RF power amplifier shall be constructed or modified during any calendar year by an amateur radio operator for use in the Amateur Radio Service without a grant of type acceptance.

(c) A list of type accepted equipment may be inspected at FCC headquarters in Washington, D.C. or at any FCC field office. Any external RF power amplifier appearing on this list as type accepted for use in the Amateur Radio Service may be marketed for use in the Amateur Radio Service.

§ 97.77 Standards for type acceptance of external radio frequency (RF) power amplifiers and external radio frequency power amplifier kits.

(a) An external radio frequency (RF) power amplifier or external RF power amplifier kit will receive a grant of type acceptance under this Part only if a grant of type acceptance would serve the public interest, convenience or necessity.

(b) To receive a grant of type acceptance under this Part, an external RF power amplifier shall meet the emission limitations of §97.73 when the amplifier is—

(1) Operated at its full output power;

(2) Placed in the "standby" or "off" positions, but still connected to the transmitter; and

(3) Driven with at least 50 watts mean radio frequency input power (unless a higher drive level is specified).

(c) To receive a grant of type acceptance under this part, an external RF power amplifier shall not be capable of operation on any frequency or frequencies between 24.00 MHz and 35.00 MHz. The amplifier will be deemed incapable of operation between 24.00 MHz and 35.00 MHz if—

(1) The amplifier has no more than 6 decibels of gain between 24.00 MHz and 26.00 MHz and between 28.00 MHz and 35.00 MHz. (This gain is determined by the ratio of the input RF driving signal (mean power measurement) to the mean RF output power of the amplifier.); and

(2) The amplifier exhibits no amplification (0 decibels of gain) between 26.00 MHz and 28.00 MHz.

(d) Type acceptance of external radio frequency power amplifiers or amplifier kits may be denied when denial serves the public interest, convenience or necessity by preventing the use of these amplifiers in services other than the Amateur Radio Service. Other uses of these amplifiers, such as in the Citizens Band Radio Service, is prohibited (Section 95.509). Examples of features which may result in dismissal or denial of an application for type acceptance of an external RF power amplifier include, but are not limited to, the following:

(1) Any accessible wiring which, when altered, would permit operation of the amplifier in a manner contrary to the FCC's Rules;

(2) Circuit boards or similar circuitry to facilitate the addition of components to change the amplifier's operating characteristics in a manner contrary to the FCC's Rules;

(3) Instructions for operation or modification of the amplifier in a manner contrary to the FCC's Rules;

(4) Any internal or external controls or adjustments to facilitate operation of the amplifier in a manner contrary to the FCC's Rules.

(5) Any internal radio frequency sensing circuitry or any external switch, the purpose of which is to place the amplifier in the transmit mode;

(6) The incorporation of more gain in the amplifier than is necessary to operate in the Amateur Radio Service. For purposes of this paragraph, an amplifier must meet the following requirements:

(i) No amplifier shall be capable of achieving designed output (or designed d.c. input) power when driven with less than 50 watts mean radio frequency input power;

(ii) No amplifier shall be capable of amplifying the input RF driving signal by more than 13 decibels. (This gain limitation is determined by the ratio of the input RF driving signal (mean power) to the mean RF output power of the amplifier). If the amplifier has a designed d.c. input power of less than 1000 watts, the gain allowance is reduced accordingly. (For example, an amplifier with a designed d.c. input power of 500 watts shall not be capable of amplifying the input RF driving signal (mean power measurement) by more than 10 decibels, compared to the mean RF output power of the amplifier.);

(iii) The amplifier shall not exhibit more gain than permitted by paragraph (d)(6)(ii) of this section when driven by a radio frequency input signal of less than 50 watts mean power; and

(iv) The amplifier shall be capable of sustained operation at its designed power level.

(7) Any attenuation in the input of the amplifier which, when removed or modified, would permit the amplifier to function at its designed output power when driven by a radio frequency input signal of less than 50 watts mean power.

SUBPART D—OPERATING REQUIREMENTS AND PROCEDURES

GENERAL

§ 97.78 Practice to be observed by all licensees.

In all respects not specifically covered by these regulations each amateur station shall be operated in accordance with good engineering and good amateur practice.

§ 97.79 Control operator requirements.

(a) The licensee of an amateur station shall be responsible for its proper operation.

(b) Every amateur radio station, when in operation, shall have a control operator at an authorized control point. The control operator shall be on duty, except where the station is operated under automatic control. The control operator may be the station licensee, if a licensed amateur radio operator, or may be another amateur radio operator with the required class of license and designated by the station licensee. The control operator shall also be responsible, together with the station licensee, for the proper operation of the station.

(c) An amateur station may only be operated in the manner and to the extent permitted by the operator privileges authorized for the class of license held by the control operator, but may exceed those of the station licensee provided proper station identification procedures are performed.

(d) The licensee of an amateur radio station may permit any third party to participate in amateur radio communication from his station, provided that a control operator is present and continuously monitors and supervises the radio communication to insure compliance with the rules.

§ 97.81 Authorized apparatus.

An amateur station license authorizes the use under control of the licensee of all transmitting apparatus at the fixed location specified in the station license which is operated on any frequency, or frequencies allocated to the amateur service, and in addition authorizes the use, under control of the licensee, of portable and mobile transmitting apparatus operated at other locations.

§ 97.82 Availability of operator license.

The original operator license of each operator shall be kept in the personal possession of the operator while operating an amateur station. When operating an amateur station at a fixed location, however, the license may be posted in a conspicuous place in the room occupied by the operator. The license shall be available for inspection by any authorized Government official whenever the operator is operating an amateur station and at other times upon request made by an authorized representative of the Commission, except when such license has been filed with application for modification or renewal thereof, or has been mutilated, lost or destroyed, and request has been made for a duplicate license in accordance with §97.57. No recognition shall be accorded to any photocopy of an operator license; however, nothing in this section shall be construed to prohibit the photocopying for other purposes of any amateur radio operator license.

§ 97.83 Availability of station license.

The original license of each amateur station or a photocopy thereof shall be posted in a conspicuous place in the room occupied by the licensed operator while the station is being operated at a fixed location or shall be kept in his personal possession. When the station is operated at other than a fixed location, the original station license or a photocopy thereof shall be kept in the personal possession of the station licensee (or a licensed representative) who shall be present at the station while it is being operated as a portable or mobile station. The original station license shall be available for inspection by any authorized Government official at all times while the station is being operated and at other times upon request made by an authorized representative of the Commission, except when such license has been filed with application for modification or renewal thereof, or has been mutilated, lost, or destroyed, and request has been made for a duplicate license in accordance with §97.57.

§ 97.84 Station identification.

(a) An amateur station shall be identified by the transmission of its call sign at the beginning and end of each single transmission or exchange of transmissions and at intervals not to exceed 10 minutes during any single transmission or exchange of transmissions of more than 10 minutes duration. Additionally, at the end of an exchange of telegraphy (other than teleprinter) or telephony transmissions between amateur stations, the call sign (or the generally accepted network identifier) shall be given for the station, or for at least one of the group of stations, with which communication was established.

(b) Under conditions when the control operator is other than the station licensee, the station identification shall be the assigned call sign for that station. However, when a station is operated within the privileges of the operator's class of license but which exceeds those of the station licensee, station identification shall be made by following the station call sign with the operator's primary station call sign (i.e. WN4XYZ/W4XX).

(c) An amateur radio station in repeater operation or a station in auxiliary operation used to relay automatically the signals of other stations in a system of stations shall be identified by radiotelephony or radiotelegraphy at a level of modulation sufficient to be intelligible through the repeated transmission at intervals not to exceed ten minutes.

(d) When an amateur radio station is in repeater or auxiliary operation, the following additional identifying information shall be transmitted:

(1) When identifying by radiotelephony, a station in repeater operation shall transmit the word "repeater" at the

end of the station call sign. When identifying by radiotelegraphy, a station in repeater operation shall transmit the fraction bar $\overline{DN}$ followed by the letters "RPT" or "R" at the end of the station call sign. (The requirements of this subparagraph do not apply to stations having call signs prefixed by the letters "WR".)

(2) When identifying by radiotelephony, a station in auxiliary operation shall transmit the word "auxiliary" at the end of the station call sign. When identifying by radiotelegraphy, a station in auxiliary operation shall transmit the fraction bar $\overline{DN}$ followed by the letters "AUX" or "A" at the end of the station call sign.

(e) A station in auxiliary operation may be identified by the call sign of its associated station.

(f) When operating under the authority of an Interim Amateur Permit with privileges authorized by the Permit, but which exceed the privileges of the licensee's permanent operator license, the station must be identified in the following manner:

(1) On radiotelephony, by the transmission of the station call sign, followed by the word "interim", followed by the special identifier shown on the Interim Permit;

(2) On radiotelegraphy, by the transmission of the station call sign, followed by the fraction bar $\overline{DN}$, followed by the special identifier shown on the interim permit.

(g) The identification required by this section shall be given on each frequency being utilized for transmission and given on each frequency being utilized for transmission and shall be transmitted either by telegraphy using the interna-
shall be transmitted either by telegraphy using the interna-
tional Morse code, or by telephony, using the English language. If the identification required by this section is made by an automatic device used only for identification by telegraphy, the code speed shall not exceed 20 words per minute. The Commission encourages the use of a nationally or internationally recognized standard phonetic alphabet as an aid for correct telephone identification.

§ 97.85 Repeater operation.

(a) Emissions from a station in repeater operation shall be discontinued within five seconds after cessation of radiocommunications by the user station. Provisions to limit automatically the access to a station in repeater operation may be incorporated but are not mandatory.

(b) Except for operation under automatic control, as provided in paragraph (e) of this section, the transmitting and receiving frequencies used by a station in repeater operation shall be continuously monitored by a control operator immediately before and during periods of operation.

(c) A station in repeater operation shall not concurrently retransmit amateur radio signals on more than one frequency in the same amateur frequency band, from the same location.

(d) A station in repeater operation shall be operated in a manner ensuring that it is not used for one-way communications, except as provided in §97.91.

(e) A station in repeater operation, either locally controlled or remotely controlled, may also be operated by automatic control when devices have been installed and procedures have been implemented to ensure compliance with the rules when a duty control operator is not present at a control point of the station. Upon notification by the Commission of improper operation of a station under automatic control, operation under automatic control shall be immediately discontinued until all deficiencies have been corrected.

§ 97.86 Auxiliary operation.

(a) A station in auxiliary operation, either locally controlled or remotely controlled, may also be operated by automatic control when it is operated as part of a system of stations in repeater operation operated under automatic control.

(b) If a station in auxiliary operation is relaying signals of another amateur radio station(s) to a station in repeater operation, the station in auxiliary operation may use an input (receiving) frequency in frequency bands reserved for auxiliary operation, repeater operation, or both.

(c) A station in auxiliary operation shall be used only to communicate with stations shown in the system network diagram.

§ 97.88 Operation of a station by remote control.

An amateur radio station may be operated by remote control only if there is compliance with the following:

(a) A photocopy of the remotely controlled station license shall be—

(1) posted in a conspicuous place at the remotely controlled transmitter location, and

(2) placed in the log of each authorized control operator.

(b) The name, address, and telephone number of the remotely controlled station licensee and at least one control operator shall be posted in a conspicuous place at the remotely controlled transmitter location.

(c) Except for operation under automatic control, a con-trol operator shall be on duty when the station is being remotely controlled. Immediately before and during the periods the remotely controlled station is in operation, the frequencies used for emission by the remotely controlled station shall be monitored by the control operator. The control operator shall terminate all transmissions upon any deviation from the rules.

(d) Provisions must be incorporated to limit transmission to a period of no more than 3 minutes in the event of malfunction in the control link.

(e) A station in repeater operation shall be operated by radio remote control only when the control link uses frequencies other than the input (receiving) frequencies of the station in repeater operation.

§ 97.89 Points of Communications.

(a) Amateur stations may communicate with:

(1) Other amateur stations, excepting those prohibited by Appendix 2.

(2) Stations in other services licensed by the Commission and with U.S Government stations for civil defense purposes in accordance with Subpart F of this part, in emergencies and, on a temporary basis, for test purposes.

(3) Any station which is authorized by the Commission to communicate with amateur stations.

(b) Amateur stations may be used for transmitting signals, or communications, or energy, to receiving apparatus for the measurement of emissions, temporary observation of transmission phenomena, radio control of remote objects, and similar experimental purposes and for the purposes set forth in §97.91.

§ 97.91 One-way communications.

In addition to the experimental one-way transmission permitted by §97.89, the following kinds of one-way communications, addressed to amateur stations, are authorized and will not be construed as broadcasting: (a) Emergency communications, including bona fide emergency drill practice transmissions; (b) Information bulletins consisting solely of subject matter having direct interest to the amateur radio service as such; (c) Round-table discussions or net-type operations where more than two amateur stations are in communication, each station taking a turn at transmitting to other station(s) of the group; and (d) Code practice transmissions intended for persons learning or improving proficiency in the international Morse code.

§ 97.93 Modulation of carrier.

Except for brief tests or adjustments, an amateur radiotelephone station shall not emit a carrier wave on frequencies below 51 megahertz unless modulated for the purpose of communication. Single audiofrequency tones may be transmitted for test purposes of short duration for the development and perfection of amateur radio telephone equipment.

Station Operation Away From Authorized Location

§ 97.95 Operation away from the authorized fixed station location.

(a) Operation within the United States, its territories or possessions is permitted as follows:

(1) When there is no change in the authorized fixed operation station location, an amateur radio station, other than a military recreation station, may be operated portable or mobile under its station license anywhere in the United States, its territories or possessions, subject to §97.61.

(2) When the authorized fixed station location is changed, the licensee shall submit an application for modification of the station license in accordance with §97.47.

(b) When outside the continental limits of the United States, its territories, or possessions, an amateur radio station may be operated as portable or mobile only under the following conditions:

(1) Operation may not be conducted within the jurisdiction of a foreign government except pursuant to, and in accordance with express authority granted to the licensee by such foreign government. When a foreign government permits Commission licensees to operate within its territory, the amateur frequency bands which may be used shall be as prescribed or limited by that government. (See Appendix 4 of this Part for the text of treaties or agreements between the United States and foreign governments relative to reciprocal amateur radio operation.)

(2) When outside the jurisdiction of a foreign government, amateur operation may be conducted within ITU Region 2 subject to the limitations of, and on those frequency bands listed in, §97.61.

(3) When outside the jurisdiction of a foreign government, amateur operation may be conducted within ITU Regions 1 and 3 on the following frequencies, subject to the limitations and provisions of Section IV of Article 5 of the Radio Regulations of the ITU:

(i)	REGION 1	REGION 3
	3.5-3.8 MHz	1.8-2.0 MHz
	7.0-7.1 MHz	3.5-3.9 MHz
	14.0-14.35 MHz	7.0-7.1 MHz
	21.0-21.45 MHz	14.0-14.35 MHz
	28.0-29.7 MHz	21.0-21.45 MHz
	144-146 MHz	28.0-29.7 MHz
	430-440 MHz	50.0-54.0 MHz
	1215-1300 MHz	144-148 MHz
	2300-2450 MHz	420-450 MHz
		1215-1300 MHz
		2300-2450 MHz

(ii) Operation on amateur frequency bands above 2450 MHz may be conducted subject to the limitations and provisions of Section IV of Article 5 of the Radio Regulations of the ITU.

(4) Except as otherwise provided, amateur operation conducted outside the jurisdiction of a foreign government shall comply with all requirements of Part 97 of this Chapter.

SPECIAL PROVISIONS

§ 97.99 Stations used only for radio control of remote model crafts and vehicles.

An amateur transmitter when used for the purpose of transmitting radio signals intended only for the control of a remote model craft or vehicle and having mean output power not exceeding one watt may be operated under the special provisions of this section provided an executed Transmitter Identification Card (FCC Form 452-C) or a plate made of a durable substance indicating the station call sign and licensee's name and address is affixed to the transmitter.

(a) Station identification is not required for transmissions directed only to a remote model craft or vehicle.

(b) Transmissions containing only control signals directed only to a remote model craft or vehicle are not considered to be codes or ciphers in the context of the meaning of §97.117.

(c) Station logs need not indicate the times of commencing and terminating each transmission or series of transmissions.

§ 97.101 Mobile stations aboard ships or aircraft.

In addition to complying with all other applicable rules, an amateur mobile station operated on board a ship or aircraft must comply with all of the following special conditions: (a) The installation and operation of the amateur mobile station shall be approved by the master of the ship or captain of the aircraft; (b) The amateur mobile station shall be separate from and independent of all other radio equipment, if any, installed on board the same ship or aircraft; (c) The electrical installation of the amateur mobile station shall be in accord with the rules applicable to ships or aircraft as promulgated by the appropriate government agency; (d) The operation of the amateur mobile station shall not interfere with the efficient operation of any other radio equipment installed on board the same ship or aircraft; and (e) The amateur mobile station and its associ-

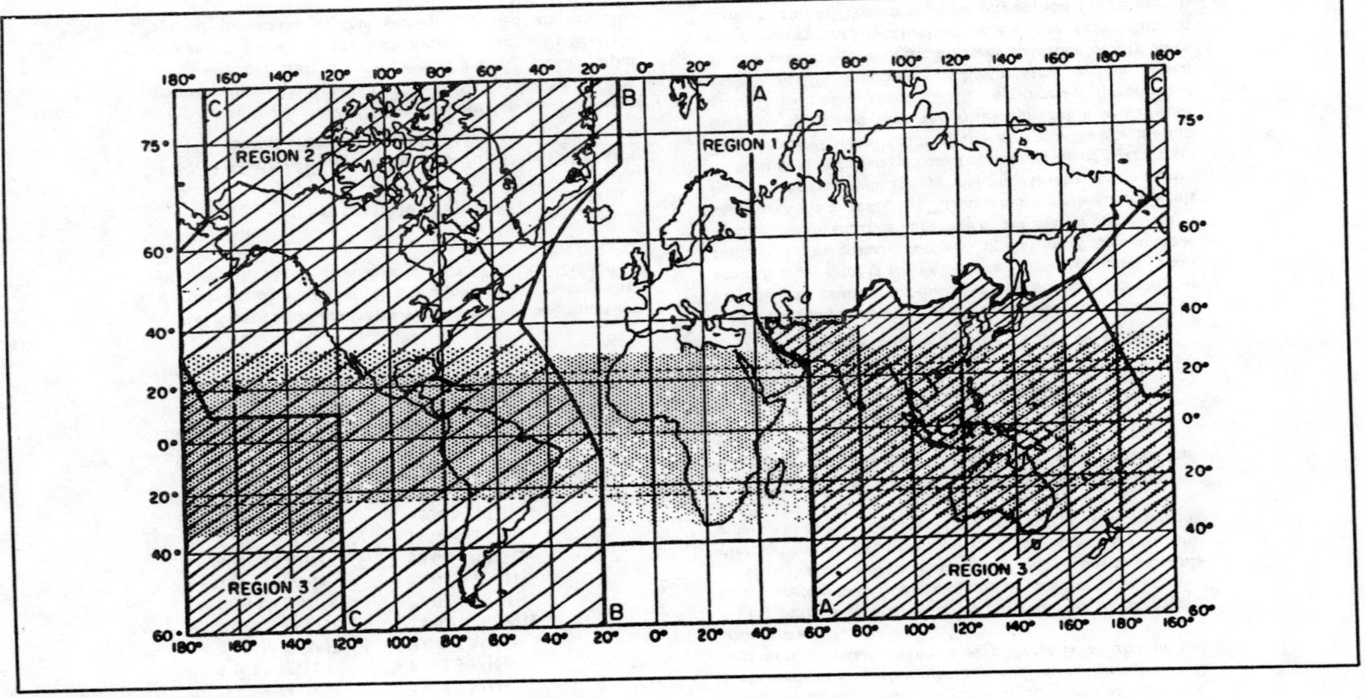

Fig. C-1. Chart of ITU Regions.

ated equipment, either in itself or in its method of operation, shall not constitute a hazard to the safety of life or property.

LOGS

§ 97.103 Station log requirements.

An accurate legible account of station operation shall be entered into a log for each amateur radio station. The following items shall be entered as a minimum:

(a) The call sign of the station, the signature of the station licensee, or a photocopy of the station license.

(b) The locations and dates upon which fixed operation of the station was initiated and terminated. If applicable, the location and dates upon which portable operation was initiated and terminated at each location.

(1) The date and time periods the duty control operator for the station was other than the station licensee, and the signature and primary station call sign of that duty control operator.

(2) A notation of third party traffic sent or received, including names of all third parties, and a brief description of the traffic content. This entry may be in a form other than written, but one which can be readily transcribed by the licensee into written form.

(3) Upon direction of the Commission, additional information as directed shall be recorded in the station log.

(c) In addition to the other information required by this section, the log of a remotely controlled station shall have entered the names, addresses, and call signs of all authorized control operators and a functional block diagram of, and a technical explanation sufficient to describe the operation of the control link. Additionally, the following information shall be entered:

(1) A description of the measures taken for protection against access to the remotely controlled station by unauthorized persons;

(2) A description of the measures taken for protection against unauthorized station operation, either through activation of the control link, or otherwise;

(3) A description of the provisions for shutting down the station in the case of control link malfunction; and

(4) A description of the means used for monitoring the transmitting frequencies.

(d) When a station has one or more associated stations, that is, stations in repeater or auxiliary operation, a system network diagram shall be entered in the station log.

(e) In addition to the other information required by this section, the log of a station in repeater operation transmitting with an effective radiated power greater than the minimum effective radiated power listed in §97.67(c) for the frequency band in use shall contain the following:

(1) The location of the station transmitting antenna, marked upon a topographic map having a scale of 1:250,000 and contour intervals[1];

(2) The antenna transmitting height above average terrain[2];

(3) The effective radiated power in the horizontal plane

---

[1]Indexes and ordering information for suitable maps are available from the U.S. Geological Survey, Washington, D.C. 20242, or from the Federal Center, Denver, Colorado 80255.

[2]See Appendix 5.

for the main lobe of the antenna pattern, calculated for maximum transmitter output power;

(4) The transmitter output power;

(5) The loss in the transmission line between the transmitter and the antenna, expressed in decibels;

(6) The relative gain in the horizontal plane of the transmitting antenna; and

(7) The horizontal and vertical radiation patterns of the transmitting antenna, with reference to true north (for horizontal pattern only), upon polar coordinate graph paper, and the method used in determining these patterns.

(f) In addition to the other information required by this section, the log of a station in auxiliary operation shall have the following information entered:

(1) A system network diagram for each system with which the station is associated;

(2) The station transmitting band(s);

(3) The transmitter input power; and

(4) If operated by remote control, the information required by paragraph (c) of this section.

(g) Notwithstanding the provisions of §97.105, the log entries required by paragraphs (c), (d), (e), and (f) of this section shall be retained in the station log as long as the information contained in those entries is accurate.

§ 97.105 Retention of logs.

The station log shall be preserved for a period of at least 1 year following the last date of entry and retained in the possession of the licensee. Copies of the log, including the sections required to be transcribed by §97.103, shall be available to the Commission for inspection.

§ 97.107 Operation in emergencies.

In the event of an emergency disrupting normally available communication facilities in any widespread area or areas, the Commission, in its discretion, may declare that a general state of communications emergency exists, designate the area or areas concerned, and specify the amateur frequency bands, or segments of such bands, for use only by amateurs participating in emergency communication within or with such affected area or areas. Amateurs desiring to request the declaration of such a state of emergency should communicate with the Commission's Engineer in Charge of the area concerned. Whenever such declaration has been made, operation of and with amateur stations in the area concerned shall be only in accordance with the requirements set forth in this section, but such requirements shall in nowise affect other normal amateur communication in the affected area when conducted on frequencies not designated for emergency operation.

(a) All transmissions within all designated amateur communications bands[1] other than communications relating directly to relief work, emergency service, or the establishment and maintenance of efficient amateur radio networks for the handling of such communications shall be suspended.

[1]The frequency 4383.8 kHz may be used by any station authorized under this part to communicate with any other station in the State of Alaska for emergency communications. No airborne operations will be permitted on this frequency. Additionally, all stations operating on this frequency must be located in or within 50 nautical miles of the State of Alaska.

Incidental calling, answering, testing or working (including casual conversations, remarks or messages) not pertinent to constructive handling of the emergency situation shall be prohibited within these bands.

(b) The Commission may designate certain amateur stations to assist in the promulgation of information relating to the declaration of a general state of communications emergency, to monitor the designated amateur emergency communications bands, and to warn non-complying stations observed to be operating in those bands. Such station, when so designated, may transmit for that purpose on any frequency or frequencies authorized to be used by that station, provided such transmissions do not interfere with essential emergency communications in progress; however, such transmissions shall preferably be made on authorized frequencies immediately adjacent to those segments of the amateur bands being cleared for the emergency. Individual transmissions for the purpose of advising other stations of the existence of the communications emergency shall refer to this section by number (§97.107) and shall specify, briefly and concisely, the date of the Commission's declaration, the area and nature of the emergency, and the amateur frequency bands or segments of such bands which constitute the amateur emergency communications bands at the time. The designated stations shall not enter into discussions with other stations beyond furnishing essential facts relative to the emergency, or acting as advisors to stations desiring to assist in the emergency, and the operators of such designated stations shall report fully to the Commission the identity of any stations failing to comply, after notice, with any of the pertinent provisions of this section.

(c) The special conditions imposed under the provisions of this section shall cease to apply only after the Commission or its authorized representative, shall have declared such general state of communications emergency to be terminated: however, nothing in this paragraph shall be deemed to prevent the Commission from modifying the terms of its declaration from time to time as may be necessary during the period of a communications emergency, or from removing those conditions with respect to any amateur frequency band or segment of such band which no longer appears essential to the conduct of the emergency communications.

SUBPART E—PROHIBITED PRACTICES AND ADMINISTRATIVE SANCTIONS

PROHIBITED TRANSMISSIONS AND PRACTICES

§ 97.112 No remuneration for use of station.

(a) An amateur station shall not be used to transmit or receive messages for hire, nor for communication for material compensation, direct or indirect, paid or promised.

(b) Control operators of a club station may be compensated when the club station is operated primarily for the purpose of conducting amateur radiocommunication to provide telegraphy practice transmissions intended for persons learning or improving proficiency in the international Morse code, or to disseminate information bulletins consisting solely of subject matter having direct interest to the Amateur Radio Service provided:

(1) The station conducts telegraphy practice and bulletin transmission for at least 40 hours per week.

(2) The station schedules operations on all allocated medium and high frequency amateur bands using reasonable measures to maximize coverage.

(3) The schedule of normal operating times and frequencies is published at least 30 days in advance of the actual transmissions.

Control operators may accept compensation only for such periods of time during which the station is transmitting telegraphy practice or bulletins. A control operator shall not accept any direct or indirect compensation for periods during which the station is transmitting material other than telegraphy practice or bulletins.

§ 97.113 Broadcasting prohibited.

Subject to the provisions of §97.91, an amateur station shall not be used to engage in any form of broadcasting, that is, the dissemination of radio communications intended to be received by the public directly or by the intermediary of relay stations, nor for the retransmission by automatic means of programs or signals emanating from any class of station other than amateur. The foregoing provisions shall not be construed to prohibit amateur operators from giving their consent to the rebroadcast by broadcast stations of the transmissions of their amateur stations, provided, that the transmissions of the amateur stations shall not contain any direct or indirect reference to the rebroadcast.

§ 97.114 Third party traffic.

The transmission or delivery of the following amateur radiocommunication is prohibited:

(a) International third party traffic except with countries which have assented thereto.

(b) Third party traffic involving material compensation, either tangible or intangible, direct or indirect, to a third party, a station licensee, a control operator, or any other person.

(c) Except for an emergency communication as defined in this part, third party traffic consisting of business communications on behalf of any party. For the purpose of this section business communication shall mean any transmission or communication the purpose of which is to facilitate the regular business or commercial affairs of any party.

§ 97.115 Music prohibited.

The transmission of music by an amateur station is forbidden.

§ 97.116 Amateur radiocommunication for unlawful purposes prohibited.

The transmission of radiocommunication or messages by an amateur radio station for any purpose, or in connection with any activity, which is contrary to Federal, State or local law is prohibited.

§ 97.117 Codes and ciphers prohibited.

The transmission by radio of messages in codes or ciphers in domestic and international communications to or between amateur stations is prohibited. All communications regardless of type of emission employed shall be in plain language except that generally recognized abbreviations established by regulation or custom and usage are permissible as are any other abbreviations or signals where the intent is not to obscure the meaning but only to facilitate communications.

§ 97.119 Obscenity, indecency, profanity.

No licensed radio operator or other person shall transmit communications containing obscene, indecent, or profane words, language, or meaning.

§ 97.121 False signals.

No licensed radio operator shall transmit false or deceptive signals or communications by radio, or any call letter or signal which has not been assigned by proper authority to the radio station he is operating.

§ 97.123 Unidentified communications.

No licensed radio operator shall transmit unidentified radio communications or signals.

§ 97.125 Interference.

No licensed radio operator shall willfully or maliciously interfere with or cause interference to any radio communication or signal.

§ 97.126 **Retransmitting radio signals.**

(a) An amateur radio station, except a station in repeater operation or auxiliary operation, shall not automatically retransmit the radio signals of other amateur radio stations.

(b) A remotely controlled station, other than a remotely controlled station in repeater operation or auxiliary operation, shall automatically retransmit only the radio signals of stations in auxiliary operation shown on the remotely controlled station's system network diagram.

§ 97.127 Damage to apparatus.

No licensed radio operator shall willfully damage, or cause or permit to be damaged, any radio apparatus or installation in any licensed radio station.

§ 97.129 Fraudulent licenses.

No licensed radio operator or other person shall obtain or attempt to obtain, or assist another to obtain or attempt to obtain, an operator license by fraudulent means.

ADMINISTRATIVE SANCTIONS

§ 97.131 Restricted operation.

(a) If the operation of an amateur station causes general interference to the reception of transmissions from stations operating in the domestic broadcast service when receivers of good engineering design including adequate selectivity characteristics are used to receive such transmission and this fact is made known to the amateur station licensee, the amateur station shall not be operated during the hours from 8 p.m. to 10:30 p.m., local time, and on Sunday for the additional period from 10:30 a.m. until 1 p.m., local time, upon the frequency or frequencies used when the interference is created.

(b) In general, such steps as may be necessary to minimize interference to stations operating in other services may be required after investigation by the Commission.

§ 97.133 Second notice of same violation.

In every case where an amateur station licensee is cited within a period of 12 consecutive months for the second violation of the provisions of §§97.61, 97.63, 97.65, 97.71, or 97.73, the station licensee, if directed to do so by the Commission, shall not operate the station and shall not permit it to be operated from 6 p.m. to 10:30 p.m., local time, until written notice has been received authorizing the resumption of full-time operation. This notice will not be issued until the licensee has reported on the results of tests which he has conducted with at least two other amateur stations at hours other than 6 p.m. to 10:30 p.m., local time. Such tests are to be made for the specific purpose of aiding the licensee in determining whether the emissions of the station are in accordance with the Commission's rules. The licensee shall report to the Commission the observations made by the cooperating amateur licensee in relation to reported violations. This report shall include a statement as to the corrective measures taken to insure compliance with the rules.

§ 97.135 Third notice of same violation.

In every case where an amateur station licensee is cited within a period of 12 consecutive months for the third violation of §97.61, 97.63, 97.65, 97.71, or 97.73, the station licensee, if directed by the Commission, shall not operate the station and shall not permit it to be operated from 8 a.m. to 12 midnight, local time, except for the purpose of transmitting a prearranged test to be observed by a monitoring station of the Commission to be designated in each particular case. The station shall not be permitted to resume operation during these hours until the licensee is authorized by the Commission, following the test, to resume full-time operation. The results of the test and the licensee's record shall be considered in determining the advisability of suspending the operator license or revoking the station license, or both.

§ 97.137 Answers to notices of violations.

Any licensee receiving official notice of a violation of the terms of the Communications Act of 1934, as amended, any legislative act, Executive order, treaty to which the United States is a party, or the rules and regulations of the Federal Communications Commission, shall, within 10 days from such receipt, send a written answer direct to the office of the Commission originating the official notice: *Provided, however,* That if an answer cannot be sent or an acknowledgement made within such 10-day period by reason of illness or other unavoidable circumstances, acknowledgement and answer shall be made at the earliest practicable date with a satisfactory explanation of the delay. The answer to each notice shall be complete in itself and shall not be abbreviated by reference to other communications or answers to other notices. If the notice relates to some violation that may be due to the physical or electrical characteristics of transmitting apparatus, the answer shall state fully what steps, if any, are taken to prevent future violations, and if any new apparatus is to be installed, the date such apparatus was ordered, the name of the manufacturer, and promised date of delivery. If the notice of violation relates to some lack of attention to or improper operation of the transmitter, the name of the operator in charge shall be given.

SUBPART F—RADIO AMATEUR CIVIL EMERGENCY SERVICE (RACES)

GENERAL

§ 97.161 Basis and purpose.

The Radio Amateur Civil Emergency Service provides for amateur radio operation for civil defense communications purposes only, during periods of local, regional or national civil emergencies, including any emergency which may necessitate invoking of the President's War Emergency Powers under the provisions of section 606 of the Communications Act of 1934, as amended.

§ 97.163 Definitions.

For the purposes of this Subpart, the following definitions are applicable:

(a) *Radio Amateur Civil Emergency Service*. A radiocommunication service conducted by volunteer licensed amateur radio operators, for providing emergency radiocommunications to local, regional, or state civil defense organizations.

(b) *RACES station*. An amateur radio station licensed to a civil defense organization, at a specific land location, for the purpose of providing the facilities for amateur radio operators to conduct amateur radiocommunications in the Radio Amateur Civil Emergency Service.

§ 97.165 Applicability of rules.

In all cases not specifically covered by the provisions contained in this Subpart, amateur radio stations and RACES stations shall be governed by the provisions of the rules governing amateur radio stations and operators (Subpart A through E of this part).

STATION AUTHORIZATIONS

§ 97.169 Station license required.

No transmitting station shall be operated in the Radio Amateur Civil Emergency Service unless:

(a) The station is licensed as a RACES station by the Federal Communications Commission, or

(b) The station is an amateur radio station licensed by the Federal Communications Commission, and is certified by the responsible civil defense organization as registered with that organization.

§ 97.171 Eligibility for RACES station license.

A RACES station will only be licensed to a local, regional, or state civil defense organization.

§ 97.173 Application for RACES station license.

(a) Each application for a RACES station license shall be made on the FCC Form 610-B.

(b) The application shall be signed by the civil defense official responsible for the coordination of all civil defense activities in the area concerned.

(c) The application shall be countersigned by the responsible official for the governmental entity served by the civil defense organization.

(d) If the application is for a RACES station to be in any special manner covered by §97.41, those showings specified for non-RACES stations shall also be submitted.

§ 97.175 Amateur radio station registration in civil defense organization.

No amateur radio station shall be operated in the Radio Amateur Civil Emergency Service unless it is certified as registered in a civil defense organization by that organization.

OPERATING REQUIREMENTS

§ 97.177 Operator requirements.

No person shall be the control operator of a RACES station, or shall be the control operator of an amateur radio station conducting communications in the Radio Amateur Civil Emergency Service unless that person holds a valid amateur radio operator license and is certified as enrolled in a civil defense organization by that organization.

§ 97.179 Operator privileges.

Operator privileges in the Radio Amateur Civil Emergency Service are dependent upon, and identical to, those for the class of operator license held in the Amateur Radio Service.

§ 97.181 Availability of RACES station license and operator licenses.

(a) The original license of each RACES station, or a photocopy thereof, shall be attached to each transmitter of such station, and at each control point of such station. Whenever a photocopy of the RACES station license is utilized in compliance with this requirement, the original station license shall be available for inspection by any authorized Government official at all times while the station is being operated and at other times upon request made by an authorized representative of the Commission, except when such license has been filed with application for modification or renewal thereof, or has been mutilated, lost, or destroyed, and request has been made for a duplicate license in accordance with §97.57.

(b) In addition to the operator license availability requirements of §97.82, a photocopy of the control operator's amateur radio operator license shall be posted at a conspicuous place at the control point for the RACES station.

TECHNICAL REQUIREMENTS

§ 97.185 Frequencies available.

(a) All of the authorized frequencies and emissions allocated to the Amateur Radio Service are also available to the Radio Amateur Civil Emergency Service on a shared basis.

(b) In the event of an emergency which necessitates the invoking of the President's War Emergency Powers under the provisions of §606 of the Communications Act of 1934 as amended, unless otherwise modified or directed, RACES stations and amateur radio stations participating in RACES will be limited in operation to the following:

FREQUENCY OR FREQUENCY BANDS	*Limitations*
kHz:	
1800-1825	1
1975-2000	1
3500-3510	

Frequency	Notes
3510-3516	4
3516-3550	2,4
3984-4000	
3997	3
7097-7103	4
7103-7125	2,4
7245-7255	2,4
14047-14053	4
14220-14230	2,4
21047-21053	4
MHz:	
28.55-28.75	
29.45-29.65	
50.35-50.75	
53.30	3
53.35-53.75	
145.17-145.71	
146.79-147.33	
220-225	5

(c) Limitations: (1) Use of frequencies in the band 1800-2000 kHz is subject to the priority of the LORAN system of radionavigation in this band and to the geographical, frequency, emission, and power limitations contained in §97.61 governing amateur radio stations and operators (Subparts A through E of this part).

(2) The availability of the frequency bands 3515-3550 kHz, 7103-7125 kHz, 7245-7247 kHz, 7253-7255 kHz, 14220-14222 kHz, and 14228-14230 kHz for use during periods of actual civil defense emergency is limited to the initial 30 days of such emergency, unless otherwise ordered by the Commission.

(3) For use in emergency areas when required to make initial contact with a military unit; also, for communications with military stations on matters requiring coordinations.

(4) For use by all authorized stations only in the continental United States, except that the bands 7245-7255 kHz and 14220-14230 kHz are also available in Alaska, Hawaii, Puerto Rico, and the Virgin Islands.

(5) Those stations operating in the band 220-225 MHz shall not cause harmful interference to the government radiolocation service.

§ 97.189 Point of communications.

(a) RACES stations may only be used to communicate with:

(1) Other RACES stations;

(2) Amateur radio stations certified as being registered with a civil defense organization, by that organization;

(3) Stations in the Disaster Communications Service;

(4) Stations of the United States Government authorized by the responsible agency to exchange communications with RACES stations;

(5) Any other station in any other service regulated by the Federal Communications Commission, whenever such station is authorized by the Commission, to exchange communications with stations in the Radio Amateur Civil Emergency Service.

(b) Amateur radio stations registered with a civil defense organization may only be used to communicate with:

(1) RACES stations licensed to the civil defense organization with which the amateur radio station is registered:

(2) Any of the following stations upon authorization of the responsible civil defense official for the organization in which the amateur radio station is registered:

(i) Any RACES station licensed to other civil defense organizations;

(ii) Amateur radio stations registered with the same or another civil defense organization;

(iii) Stations in the Disaster Communications Service;

(iv) Stations of the United States Government authorized by the responsible agency to exchange communications with RACES stations;

(v) Any other station in any other service regulated by the Federal Communications Commission, whenever such station is authorized by the Commission to exchange communications with stations in the Radio Amateur Civil Emergency Service.

§ 97.191 Permissible communications.

All communications in the Radio Amateur Civil Emergency Service must be specifically authorized by the civil defense organization for the area served. Stations in this service may transmit only civil defense communications of the following types:

(a) Communications concerning impending or actual conditions jeopardizing the public safety, or affecting the national defense or security during periods of local, regional, or national civil emergencies:

(1) Communications directly concerning the immediate safety of life or individuals, the immediate protection of property, maintenance of law and order, alleviation of human suffering and need, and the combating of armed attack or sabotage;

(2) Communications directly concerning the accumulation and dissemination of public information or instructions to the civilian population essential to the activities of the civil defense organization or other authorized governmental or relief agencies.

(b) Communications for training drills and tests necessary to ensure the establishment and maintenance of orderly and efficient operation of the Radio Amateur Civil Emergency Service as ordered by the responsible civil defense organization served. Such tests and drills may not exceed a total time of one hour per week.

(c) Brief one way transmissions for the testing and adjustment of equipment.

§ 97.193 Limitations on the use of RACES stations.

(a) No station in the Radio Amateur Civil Emergency Service shall be used to transmit or to receive messages for hire, nor for communications for material compensation, direct or indirect, paid or promised.

(b) All messages which are transmitted in connection with drills or tests shall be clearly identified as such by use of the words "drill" or "test", as appropriate, in the body of the messages.

SUBPART G—OPERATION OF AMATEUR RADIO STATIONS IN THE UNITED STATES BY ALIENS PURSUANT TO RECIPROCAL AGREEMENTS

§ 97.301 Basis, purpose, and scope.

(a) The rules in this subpart are based on, and are applicable solely to, alien amateur operations pursuant to section 303(1)(3) and 310(a) of the Communications Act of 1934, as amended. (See Pub. L. 93-505, 88 Stat. 1576.)

(b) The purpose of this subpart is to implement Public Law 88-313 by prescribing the rules under which an alien, who holds an amateur operator and station license issued by his government (hereafter referred to as an alien amateur), may operate an amateur radio station in the United States, in its possessions, and in the Commonwealth of Puerto Rico (hereafter referred to only as the United States).

§ 97.303 Permit required.

(a) Before he may operate an amateur radio station in the United States, under the provisions of sections 303(1)(2) and 310(a) of the Communications Act of 1934, as amended, an alien amateur licensee must obtain a permit for such operation from the Federal Communications Commission. A permit for such operation shall be issued only to an alien holding a valid amateur operator and station authorization from his government, and only when there is in effect a bilateral agreement between the United States and that government for such operation on a reciprocal basis by United States amateur radio operators.

§ 97.305 Application for permit.

(a) Application for a permit shall be made on FCC Form 610-A. Form 610-A may be obtained from the Commission's Washington, D.C., office, from any of the Commission's field offices and, in some instances, from United States missions abroad.

(b) The application form shall be completed in full in English and signed by the applicant. A photocopy of the applicant's amateur operator and station license issued by his government shall be filed with the application. The Commission may require the applicant to furnish additional information. The application must be filed by mail or in person with the Federal Communications Commission, Gettysburg, Pennsylvania 17325, U.S.A. To allow sufficient time for processing, the application should be filed at least 60 days before the date on which the applicant desires to commence operation.

§ 97.307 Issuance of permit.

(a) The Commission may issue a permit to an alien amateur under such terms and conditions as it deems appropriate. If a change in the terms of a permit is desired, an application for modification of the permit is required. If operation beyond the expiration date of a permit is desired, an application for renewal of the permit is required. In any case in which the permittee has, in accordance with the provisions of this subpart, made a timely and sufficient application for renewal of an unexpired permit, such permit shall not expire until the application has been finally determined. Applications for modification or for renewal of a permit shall be filed on FCC Form 610-A.

(b) The Commission, in its discretion may deny any application for a permit under this subpart. If an application is denied, the applicant will be notified by letter. The applicant may, within 90 days of the mailing of such letter, request the Commission to reconsider its action.

(c) Normally, a permit will be issued to expire 1 year after issuance but in no event after the expiration of the license issued to the alien amateur by his government.

§ 97.309 Modification, suspension, or cancellation of permit.

At any time the Commission may, in its discretion, modify, suspend, or cancel any permit issued under this subpart. In this event, the permittee will be notified of the Commission's action by letter mailed to his mailing address in the United States and the permittee shall comply immediately. A permittee may, within 90 days of the mailing of such letter, request the Commission to reconsider its action. The filing of a request for reconsideration shall not stay the effectiveness of that action, but the Commission may stay its action on its own motion.

§ 97.311 Operating conditions.

(a) The alien amateur may not under any circumstances begin operation until he has received a permit issued by the Commission.

(b) Operation of an amateur station by an alien amateur under a permit issued by the Commission must comply with all of the following:

(1) The terms of the bilateral agreement between the alien amateur's government and the government of the United States;

(2) The provisions of this subpart and of Subparts A through E of this part;

(3) The operating terms and conditions of the license issued to the alien amateur by his government; and

(4) Any further conditions specified on the permit issued by the Commission.

§ 97.313 Station identification.

(a) The alien amateur shall identify his station as follows:

(1) Radio telegraph operation: The amateur shall transmit the call sign issued to him by the licensing country followed by a slant (/) sign and the United States amateur call sign prefix letter(s) and number appropriate to the location of his station.

(2) Radiotelephone operation: The amateur shall transmit the call sign issued to him by the licensing country followed by the words "fixed", "portable" or "mobile", as appropriate, and the United States amateur call sign prefix letter(s) and number appropriate to the location of his station. The identification shall be made in the English language.

(b) At least once during each contact with another amateur station, the alien amateur shall indicate, in English, the geographical location of his station as nearly as possible by city and state, commonwealth, or possession.

SUBPART H—(RESERVED)

APPENDICES

APPENDIX 1

EXAMINATION POINTS

Examinations for amateur radio operator licenses are conducted at the Commission's office in Washington, D.C., and at each field office of the Commission on the days designated by the Engineer in Charge of each office. Specific dates should be obtained from the Engineer in Charge of the nearest field office of the Commission.

Examinations are also given at prescribed intervals in the cities listed in the Commission's current Examination Schedule, copies of which are available from the Federal Communications Commission Regional Services Division, Washington, D.C. 20554, or from any one of the Commission's field offices listed in §0.121.

APPENDIX 2

Extracts From Radio Regulations Annexed to the International Telecommunication Convention (Geneva, 1959)

ARTICLE 41—AMATEUR STATIONS

SECTION 1. Radiocommunications between amateur stations of different countries[1] shall be forbidden if the administration of one of the countries concerned has notified that it objects to such radiocommunications.

SEC. 2.(1) When transmissions between amateur stations of different countries are permitted, they shall be made in plain language and shall be limited to messages of a technical nature relating to tests and to remarks of a personal character for which, by reason of their unimportance, recourse to the public telecommunications service is not justified. It is absolutely forbidden for amateur stations to be used for transmitting international communications on behalf of third parties.

(2) The preceding provisions may be modified by special arrangements between the administration of the countries concerned.

SEC. 3(1) Any person operating the apparatus of an amateur station shall have proved that he is able to send correctly by hand and to receive correctly by ear, texts in Morse code signals. Administrations concerned may, however, waive this requirement in the case of stations making use exclusively of frequencies above 144 MHz.

(2) Administrations shall take such measures as they judge necessary to verify the technical qualifications of any person operating the apparatus of an amateur station.

SEC. 4. The maximum power of amateur stations shall be fixed by the administrations concerned, having regard to the technical qualifications of the operators and to the conditions under which these stations are to work.

SEC. 5. (1) All the general rules of the Convention and of these Regulations shall apply to amateur stations. In particular, the emitted frequency shall be as stable and as free from spurious emissions as the state of technical development for such stations permits.

(2) During the course of their transmissions, amateur stations shall transmit their call sign at short intervals.

RESOLUTION NO. 10

Relating to the use of the bands 7000 to 7100 kHz and 7100 to 7300 kHz by the Amateur Service and the Broadcasting Service.

The Administrative Radio Conference Geneva, 1959,

Considering—

(a) That the sharing of frequency bands by amateur, fixed, and broadcasting services is undesirable and should be avoided;

(b) That it is desirable to have worldwide exclusive allocations for these services in Band 7;

(c) That the band 7000 to 7100 kHz is allocated on a worldwide basis exclusively to the amateur service;

(d) That the band 7100 to 7300 kHz is allocated in Regions 1 and 3 to the broadcasting service and in Region 2 to the amateur service;

resolves,

that the broadcasting service should be prohibited from the band 7000 to 7100 kHz and that broadcasting stations operating on frequencies in this band should cease such operation;

and noting,

the provisions of No. 117 of the Radio Regulations;

further resolves,

that interregional amateur contacts should be only in the band 7000 to 7100 kHz and that the administrations should make every effort to ensure that the broadcasting service in the band 7100 to 7300 kHz, in Regions 1 and 3, does not cause interference to the amateur service in Region 2; such being consistent with the provisions of No. 117 of the Radio Regulations.

APPENDIX 3

CLASSIFICATION OF EMISSIONS

For convenient reference the tabulation below is extracted from the classification of typical emissions in Part 2 of the Commission's Rules and Regulations and in the Radio Regulations, Geneva, 1959, and it includes only those general classifications which appear most applicable to the Amateur Radio Service.

Type of modulation	Type of transmission	Symbol
Amplitude	With no modulation	A0
	Telegraph without the use of modulating audio frequency (by on-off keying).	A1
	Telegraphy by the on-off keying of an amplitude modulating audio frequency or audio frequencies or by the on-off keying of the modulated emission (special case; an unkeyed emission amplitude modulated).	A2
	Telephony	A3[1]
	Facsimile	A4
	Television	A5
Frequency (or phase).	Telegraphy by frequency shift keying without the use of a modulating audio frequency.	F1
	Telegraphy by the on-off keying of a frequency modulating audio frequency or by the on-off	F2

keying of frequency modulated emission (special case; an unkeyed emission frequency modulated). ———— F3
Telephony ———— F4
Facsimile ———— F5
Television ———— P
Pulse ————

[1](In Part 97) Unless specified otherwise, A3 includes single and double sideband with full, reduced, or suppressed carrier.

APPENDIX 4

Convention Between the United States of America and Canada, Relating to the Operation by Citizens of Either Country of Certain Radio Equipment or Stations in the Other Country (Effective May 15, 1952)

ARTICLE III

It is agreed that persons holding appropriate amateur licenses issued by either country may operate their amateur stations in the territory of the other country under the following conditions:

(a) Each visiting amateur may be required to register and receive a permit before operating any amateur station licensed by his government.

(b) The visiting amateur will identify his station by:

(1) *Radiotelegraphy operation.* The amateur call sign issued to him by the licensing country followed by a slant (/) sign and the amateur call sign prefix and call area number of the country he is visiting.

(2) *Radiotelephone operation.* The amateur call sign in English issued to him by the licensing country followed by the words, "fixed" "portable" or "mobile," as appropriate, and the amateur call sign prefix and call area number of the country he is visiting.

(c) Each amateur station shall indicate at least once during each contact with another station its geographical location as nearly as possible by city and state or city and province.

(d) In other respects the amateur station shall be operated in accordance with the laws and regulations of the country in which the station is temporarily located.

APPENDIX 5

DETERMINATION OF ANTENNA HEIGHT ABOVE AVERAGE TERRAIN

The effective height of the transmitting antenna shall be the height of the antenna's center of radiation above "average terrain." For this purpose "effective height" shall be established as follows:

(a) On a U.S. Geological Survey Map having a scale of 1:250,000, lay out eight evenly spaced radials, extending from the transmitter site to a distance of 10 miles and beginning at (0°, 45°, 90°, 135°, 180°, 225°, 270°, 315°T.) If preferred, maps of greater scale may be used.

(b) By reference to the map contour lines, establish the ground elevation above mean sea level (AMSL) at 2, 4, 6, 8, and 10 miles from the antenna structure along each radial. If no elevation figure or contour line exists for any particular point, the nearest contour line elevation shall be employed.

(c) Calculate the arithmetic average of these 40 points of elevation (5 points of each of 8 radials).

(d) The height above average terrain of the antenna is thus the height AMSL of the antenna's center of radiation, minus the height of average terrain as calculated above.

NOTE 1: Where the transmitter is located near a large body of water, certain points of established elevation may fall over water. Where it is expected that service would be provided to land areas beyond the body of water, the points at water level in that direction should be included in the calculation of average elevation. Where it is expected that service would not be provided to land areas beyond the body of water, the points at water level should not be included in the average.

NOTE 2: In instances in which this procedure might provide unreasonable figures due to the unusual nature of the local terrain, applicant may provide additional data at his own discretion, and such data may be considered if deemed significant.

---

[1]As may appear in public notices issued by the Commission.

WORD INDEX TO PART 97

A

B

C

O

P

Q

R

S

T

Index

Index